# FLOOD PLAIN MANAGEMENT  *Iowa's Experience*

# FLOOD PLAIN
## IOWA'S

~~~~~~~~~~~~~~~~~~~~~~~~~~~~~~~~~~~~~~~~~~~~~~~~~~~

THE IOWA STATE UNIVERSITY PRESS,

# MANAGEMENT
## EXPERIENCE

PAPERS PRESENTED AT THE CONFERENCE
ON FLOOD PLAIN MANAGEMENT, SIXTH WATER
RESOURCES DESIGN CONFERENCE,
IOWA STATE UNIVERSITY

*Edited by*

## MERWIN D. DOUGAL
*Assistant Professor of Civil Engineering*

AMES, IOWA, U.S.A.

St. Mary's College Library
Winona, Minnesota

## SPONSORS OF THE CONFERENCE

Iowa State University and Iowa State Water Resources Research Institute
Civil Engineering Department
University Extension
Engineering Extension

Iowa Development Commission
Planning Division

Iowa Natural Resources Council

Iowa Planning Association

Iowa Section, American Society of Agricultural Engineers

Iowa Section, American Society of Civil Engineers

### SPONSOR REPRESENTATIVES ON PLANNING COMMITTEE

Merwin D. Dougal, Iowa State University
Burl A. Parks, Iowa State University
William M. McLaughlin, Iowa Development Commission
Clifford E. Peterson, Iowa Natural Resources Council
James F. Cooper, Iowa Natural Resources Council
Jim L. Maynard, Iowa Planning Association
Jerry R. Wehrspahn, American Society of Agricultural Engineers, Iowa Section
Sulo W. Wiitala, American Society of Civil Engineers,
Iowa Section, Sanitary and Hydraulics Committee

© 1969 The Iowa State University Press, Ames, Iowa 50010. All rights reserved. Composed and printed by The Iowa State University Press. First edition, 1969. Standard Book Number: 8138–0595–3. Library of Congress Catalog Card Number: 75–83320

**THIS VOLUME** is gratefully dedicated to the meteorologists, hydrologists, engineers, and technicians of the weather and water agencies involved in flood operations. These individuals toil willingly day and night obtaining and analyzing weather information, river stage and discharge measurements, and flood inundation data. They provide flood warnings to communities and assist in evacuation and emergency measures. With unrelenting effort they are striving to alleviate the suffering and losses due to floods. They recognize the urgent need for nationwide regulation of flood plain use to prevent additional and unnecessary flood problems from occurring.

# CONTENTS

Foreword, ix
Preface, xiii
Acknowledgment, xvii
Introduction, *Paul E. Morgan,* xix

**PART 1. THE NEED FOR FLOOD PLAIN MANAGEMENT**
1. An Introductory Philosophy of Flood Plain Management, *J. W. Howe,* 3
2. Man Should Manage the Flood Plains, *James E. Goddard,* 11

**PART 2. FLOODS AND THE FLOOD POTENTIAL**
3. Floods in Iowa, *Harlan H. Schwob,* 27
4. The Flood Potential and Future Flood Problems, *J. Wm. Funk,* 37

**PART 3. ELEMENTS OF A FLOOD PLAIN MANAGEMENT PROGRAM**
5. Techniques for Developing a Comprehensive Program for Flood Plain Management, *Merwin D. Dougal,* 53

**PART 4. LAND-USE PLANNING AND DEVELOPMENT METHODS FOR FLOOD PLAIN AREAS**
6. A Brief Background of the Planning Process, *Burl A. Parks,* 79
7. Land-Use Planning for the Flood Plain, *William R. Klatt,* 85
8. Regional Consideration of Flood Plain Planning, *William S. Luhman,* 93
9. Coordination of Urban Planning and Flood Plain Development, *Eugene O. Johnson,* 103
10. Federal Programs for Urban Planning and Financing Assistance in Flood Plain Areas, *Phillip L. Larson,* 113

11. The Interaction of Urban Redevelopment and Flood Plain Management in Waterloo, Iowa, *John R. Sheaffer,* 123
12. The Role of Open Spaces in Flood Plain Management, *Donald K. Gardner,* 137

**PART 5. DIRECT CONTROL OF LAND USE THROUGH LAWS AND ORDINANCES**

13. Coordination of Planning with Regulatory Controls—An Introduction, *William M. McLaughlin,* 151
14. State Statutory Responsibility in Flood Plain Management, *Clifford E. Peterson,* 157
15. Model Flood Plain Regulations for Iowa—A Progress Report, *David J. Blair,* 167
16. Building Codes and Local Zoning and Subdivision Control for Flood Plain Management, *Jim L. Maynard,* 183

**PART 6. AGENCY FUNCTIONS AND SERVICES FOR FLOOD PLAIN MANAGEMENT ASSISTANCE**

17. Flood Plain Mapping by the U.S. Geological Survey, *D. W. Ellis,* 197
18. Assistance through the Flood Plain Management Services of the Corps of Engineers, *John N. Stephenson,* 207
19. The Iowa Flood Plain Management Program, *James F. Cooper,* 219

**PART 7. SUCCESS OF FLOOD PLAIN MANAGEMENT PROGRAMS IN OTHER STATES**

20. Lincoln's Experience in Regulating Flood Plain Development, *James H. Schroeder,* 237
21. Experience of the Tennessee Valley Authority in Local Flood Relations, *John W. Weathers,* 247

Appendix, 255

Index, 265

# FOREWORD

TWICE PREVIOUSLY, specific and concentrated attention has been directed toward vital water issues in Iowa. The first occasion followed the drought years which commenced in 1953 in southern and southwestern Iowa and culminated in a 1955 conference at which water quantity, including adequacy of supplies and allocation of "surplus" waters, was the major theme. The proceedings of this conference were published in 1956 as a book entitled *Iowa's Water Resources: Source, Uses, and Laws.*

Ten years later in 1965 water quality was the major problem confronting Iowa as well as the nation. Legislation at both the state and federal levels brought to the forefront the seriousness of water pollution control. Again an interdisciplinary conference was fostered by water resources specialists at Iowa State University and the University of Iowa. The exchange of information among the various educational, professional, and technical groups, all in a spirit of cooperation, established a framework for evaluating water pollution problems and outlined the several methods for obtaining cleaner waters for future use. Once again the proceedings were published as a book, completed in 1967 and entitled *Water Pollution Control and Abatement.*

Through these publications the valuable information presented at the two specialty conferences was made available to all who have an interest in or concern for solving the multifaceted water problems which have arisen in Iowa. The wise use and conservation of our water and related land resources demand such sharing of knowledge.

Study and research of Iowa's water problems continue at a rapid pace. An annual series of Iowa Water Resources Design Conferences was initiated in 1963 through the Engineering Extension Service by Professor Merwin Dougal of the Department of Civil Engineering. The specific objective was to coordinate the exchange of information among researchers, teachers, and the practicing profession engaged in the design of water resources projects. These conferences have been technical but interdisciplinary in scope and directed to the solution of the various water problems in Iowa. During this same period the Iowa State Water Resources Research Institute was established under the direction of Dr. Don Kirkham of Iowa State University. It is guided by a cooperative group composed of faculty representatives of the two state universities and has coordinated research efforts in water resources, including the administration of the federal funds received from the Office of Water Resources Research, Department of the Interior. Paramount to the solution of the state's water problems is an exchange of information among the institute and the several state agencies and professional societies concerning needs and priorities for research, study, and application of new ideas.

All the foregoing had a bearing upon the selection of "Flood Plain Management" as the topic for the Sixth Water Resources Design Conference in January, 1968. Also, as in the case of water pollution, increased federal and state involvement in flood plain activities had resulted from the continuing encroachment on flood-prone lands by an urbanizing nation. In addition, flood losses during the 1960's had been especially severe in Iowa. Better planning, more intensive engineering studies, and additional legal and socioeconomic opportunities were all involved in the concentrated attack needed to solve the problem of flood plain encroachment and mounting flood losses.

The sponsors and planning committee were selected early in 1967 to coordinate and plan an interdisciplinary approach. Through their intensive efforts an exceptionally strong program was developed. At the three-day conference, January 23–25, 1968, a total of 21 papers was presented by state and national leaders in the various disciplines, all working arduously to achieve a sound and effective flood plain management program for Iowa. Professor Merwin Dougal, Department of Civil Engineering, and Professor Burl Parks, Department of Landscape Architecture, guided the project for the Colleges of Engineering and Agriculture. Assistant Dean of Engineering Paul

## FOREWORD

Morgan was official host for Iowa State University, and in his welcome address he stressed the accomplishments being achieved through the several cooperative interdisciplinary programs at Iowa universities. These include not only water resources but also the field of biomedical engineering, consortia of universities to develop foreign educational institutions, and potential areas of cooperation in environmental living and world food production.

Success of these interdisciplinary conferences can be measured by the number in attendance and the quality of the papers presented for discussion. The flood plain management conference was attended by 120 specialists in the various disciplines and included representatives of local, state, and federal agencies and groups. The numerous requests received to date for information presented at the conference have indicated that the papers were pertinent to the subject.

Therefore, it appears appropriate to publish the proceedings of this important conference to complement the two previous volumes. The title, *Flood Plain Management: Iowa's Experience,* appropriately recognizes the contributions which will lead to the solution of a third aspect of Iowa's water problems. The Colleges of Engineering and Agriculture and the Iowa State Water Resources Research Institute are pleased to assist in the publication of this book. It is hoped that Iowans and others will find it of value in their efforts to solve flood problems and obtain wise and prudent use of the flood plains.

GEORGE R. TOWN
Dean, College of Engineering
Iowa State University
Ames, Iowa

FLOYD ANDRE
Dean, College of Agriculture
Iowa State University
Ames, Iowa

DON KIRKHAM
Director, Iowa State Water Resources Research Institute
Ames, Iowa

# PREFACE

> The nation needs a broader and more unified national program for managing flood losses. Flood protection has been immensely helpful in many parts of the country—and must be continued. Beyond this, additional tools and integrated policies are required to promote sound and economic development of the flood plains.

THE TASK FORCE on Federal Flood Control Policy made this opening statement in its report to Congress in 1966. This call for a policy change followed 30 years of national effort in the construction of flood protection works designed specifically to reduce the nation's flood losses. For in the Flood Control Act of 1936 Congress launched its broad flood control program, relying primarily upon engineering works both in the tributary watersheds and in the large river valleys. However, the rapid urban expansion and more intensive agricultural practices which followed World War II expedited new flood plain development. Soon the leaders in the flood plain management movement recognized and publicized the nature of the losing battle. As rapidly as flood protection works alleviated the flood problem at one location, new developments elsewhere canceled the contemplated reduction in national flood losses.

Dr. Gilbert F. White, Department of Geography, University of Chicago, and his dedicated group of researchers gained national attention as they pursued the problems of increased urban occupancy of the flood plains. The Tennessee Valley Authority, through its local flood relations program led by James E. Goddard, pioneered a federal-state-local program of flood plain regulation in an area noted for its multipurpose water control dams and reservoirs. These early efforts encouraged Congress to review its construction program, and in 1960 a flood plain information program was initiated. But the increased pace of urbanization and lack of comprehensive flood plain planning led to the 1966 study and report. A more vital program is proposed which will involve all the disciplines—technical, socioeconomic, and legal—and which concentrates on local government involvement, with state coordination and federal support.

Iowa is in a unique position which should permit it to move ahead rapidly with an expanded flood plain management program. Flood plain regulation was included in the organic act which in 1949 created the Iowa Natural Resources Council. The state regulatory role concerning flood plain development and construction has been strengthened by subsequent amendments and court rulings. More recent amendments in 1965 and 1967 further recognized the need and necessity of involving local planning and zoning agencies in the flood plain management program. Therefore, the state of Iowa today has provided adequate authority for achieving a sound program of flood plain management.

However, authority by itself does not accomplish results. Success of the Iowa flood plain management program depends upon three factors. First, technical coordination must be achieved not only across the local-state-federal hierarchy but also across the spectrum of disciplines involved—planners, engineers, attorneys, economists, educators, scientists, financing agencies, administrators, and public officials of the communities and counties in Iowa. Second, comprehensive planning of flood plains is a clear prerequisite to successful control and regulatory phases of flood plain management. The relatively nonurbanized and unoccupied lands in cities and in counties deserve immediate and rapid consideration; urban areas already developed present nonconforming use problems which will deserve detailed and concentrated attention. Third, local people must become involved in the planning process, and local acceptance of flood plain management including the regulation concept must be achieved. There is a great need for convincing property owners, developers, realtors, and public officials that flood plain management programs are of benefit to them as well as to the public. Community desire for guided growth should insist on such participation by those directly involved in flood plain development and use.

# PREFACE

This seven-part volume brings together in one integrated text the available information concerning the many phases of flood plain management. Much more will be needed before the task is completed, but it is designed to show that methods are available for beginning positive programs today. Each of the chapters is devoted to a specific phase or aspect of the subject. In all, they cover the many problems which arise in enacting local action programs for achieving optimum and wise use of the flood plains.

Part I presents the philosophy and need for flood plain management and leaves little justification or reason for not accepting public regulatory measures.

Part II presents the physical flood problem and the future flood potential in Iowa. Recognition of the full potential is necessary in evaluating the hazard in Iowa communities.

Part III is devoted to the development of action programs for flood plain management; the various elements are listed and discussed.

Part IV deals with the planning role and its many facets. Included are concepts and methods, compatible land uses, regional and city planning and zoning coordination problems, federal assistance programs in planning and urban renewal, and open-space possibilities.

Part V presents the legal aspects of flood plain management. Included are Iowa programs in planning and zoning, state statutory control, model flood plain zoning regulations, subdivision regulations, and building code provisions.

Part VI is devoted to technical flood plain information studies upon which communities can base their planning and regulatory controls.

Part VII concludes the text with two reports on the success of management programs in other states.

The Appendix includes a list of all cities in Iowa and a preliminary evaluation of the flood problems and availability of information.

This volume provides an extensive inventory of the problems and potentialities encountered to date in flood plain management. Additional information and newer methods will be needed in the future to provide for stronger programs and to cope with problems which will surely emerge. Thus a challenge remains for continued research in this important area of water resources.

MERWIN D. DOUGAL
Iowa State University

# ACKNOWLEDGMENT

THIS CONFERENCE was sponsored by the university and engineering extension service in cooperation with the State Department of Public Instruction, Division of Vocational Education.

Subvention for publication of this book was paid from funds provided by the United States Department of the Interior as authorized under the Water Resources Research Act of 1964, Public Law 88-379.

Editorial assistance was furnished in part through a grant from Iowa Community Services under Title I of the Higher Education Act of 1965.

Special recognition is given the several agencies and individuals involved in the flood plain management conference and in publishing the proceedings. Members of the 1967–1968 sanitary and hydraulics committee of the Iowa Section, American Society of Civil Engineers, assisted in much of the early planning. The time and effort of each author as well as the cooperation and assistance of his respective employer, firm, or agency were necessary in obtaining the desired expertise and content for each assigned topic. The timely assistance of Mrs. Sharon Coleman, who typed the revised and edited copy into manuscript form, is gratefully acknowledged. The assistance of Mrs. Nancy Schworm, of the Iowa State University Press editorial staff, in the final editing is appreciated, and the readability of the book is much improved through her efforts.

MERWIN D. DOUGAL

# INTRODUCTION
## Paul E. Morgan

IN HOLDING THIS sixth annual Water Resources Design Conference, we at Iowa State University wished to convey our concern, interest, and responsibility for developing sound programs in continuing education similar to this one. The need is great for *cooperation* among the various educational, professional, technical, and governmental groups of our society, which will permit us to find solutions to the complex problems of today and tomorrow.

Those attending this three-day conference were among the leaders who recognize need for cooperation in the water resources field. The program crossed governmental, educational, professional, and technical boundaries to include the following organizations and technical personnel:

Soil Conservation Service, U.S. Department of Agriculture
Geological Survey, U.S. Department of the Interior
Tennessee Valley Authority
Corps of Engineers, U.S. Army
U.S. Department of Housing and Urban Development
Development Commission, Planning Division
Natural Resources Council

PAUL E. MORGAN is Assistant Dean, College of Engineering, and Professor, Department of Civil Engineering, Iowa State University.

Educational institutions
Professional engineering societies—ASCE, ASAgE
Iowa Planning Association
Urban planning and zoning personnel
Parks and public property personnel
Consulting engineers, hydrologists, planners
Legal authorities

In addition, an exceptional amount of cooperation was shown by the varied sponsorship of Iowa State University, the Iowa Development Commission, the Iowa Natural Resources Council, the Iowa Planning Association, and the Iowa Sections of the American Societies of Agricultural Engineers and Civil Engineers. Through their efforts these several groups will assist all the disciplines in solving the many problems associated with the encroachment, use, and management of flood plains.

At Iowa State University the various people interested in water resources have recognized that cooperation is necessary to accomplish their basic educational goals in this expanding field. The Graduate College and the Water Resources Advisory Committee of Iowa State University have formulated and implemented a cooperative water resources curriculum and program. The departments of Agricultural Engineering, Agronomy, Bacteriology, Civil Engineering, Dairy and Food Industries, Economics, Entomology, Forestry, Geology, Zoology, and Household Equipment are participating in this program and offer the M.S. and Ph.D. degrees in Water Resources. This program is quite flexible and permits the entry of qualified students from any of the above departments.

Not only is there increasing cooperation among departments within the university but universities in Iowa are cooperating with each other as well. Biomedical Engineering at Iowa State is a good example of this. The Department of Electrical Engineering and the College of Veterinary Medicine at Iowa State University and the College of Medicine at the University of Iowa have a joint program in this field. This is considered by many to be outstanding and future expansion is hoped for. We at Iowa State look for additional cooperative programs to be developed among our sister institutions as the need arises.

In addition to the cooperative efforts of Iowa universities to find solutions to unique and varied problems, consortia of American universities are now forming to bring together specialized talents

## INTRODUCTION

from a wide segment of our educational resources. Iowa State University belongs to two of these consortia that are cooperating in solving many of our common problems in education and research and are assisting in the development of foreign educational institutions. The College of Engineering is presently cooperating with other members of these consortia to help in the development of the National Engineering University in Lima, Peru, the University of the Philippines in Manila, and Mindanao State University, Mindanao, Philippines.

Additional areas of cooperation undoubtedly will be necessary in the future to develop other interdisciplinary programs. Two such areas gaining prominent attention today are environmental living and world food production. It is suggested that the environmental living program might include interdepartmental cooperation among the departments of Architecture, Civil Engineering, Chemical Engineering, Economics, Government, Landscape Architecture, Water Resources, and others interested in solving the problems confronting our metropolitan areas today.

The Director General for Food and Agriculture of the World Health Organization recently stated:

> Unless we solve the problem of matching our numbers and our food production, we face chaos in the form *not only* of hunger, of poverty, fear and disease, but of bitterness, conflict and violence that will overflow every frontier. We are, in fact, already heading into it.

It appears that one method with considerable potential for success in increasing the food production capability of the world would be to develop a cooperative world food production program. This would include at Iowa State the people interested in water resources, agronomy, animal science, veterinary medicine, nutrition, chemical engineering, civil engineering, meteorology, hydrology, economics, and international law. We look for such a group to organize in the very near future, and it appears that the state of Iowa and Iowa State University could serve as a nucleus for such an important interdisciplinary program.

This water resources design group is to be commended for leading the professional world in cooperative interinstitutional, professional, and technical programs to help solve some of these broad-based worldwide problems in today's society.

Little Sioux River at Spencer, 1953. Courtesy Spencer **Daily Reporter**.

# Part 1

*Failure to recognize that the natural function of a flood plain is to carry away excess water in time of flood often has led to rapid and haphazard development on flood plains with a consequent increase in flood hazards.*

S. W. Wiitala, K. R. Jetter, and A. J. Sommerville—USGS

# THE NEED FOR FLOOD PLAIN MANAGEMENT

# 1

# AN INTRODUCTORY PHILOSOPHY
# OF FLOOD PLAIN MANAGEMENT

J. W. Howe

MANY RELEVANT POINTS should be covered in a thorough discussion of flood plain management. Inasmuch as the following chapters are devoted to specific aspects of the interdisciplinary problems which arise, it would appear appropriate to present a brief review of the whole dilemma of flood plain occupancy. That these occupants sooner or later involve the public and its elected and appointed administrators is clearly evident from the flood experience of recent years in Iowa.

The trials and tribulations which are encountered without fail in resolving this dilemma are presented here. The aim is to provide an initial insight for those desiring to implement the ideas and concepts which are the subject of this book.

## THE NATURE OF THE DILEMMA

The fact is frequently emphasized that although we have spent billions of dollars on flood protection works, the annual damage attributed to floods continues to rise.[8,9] The explanation for this para-

J. W. HOWE is Professor and Chairman, Department of Mechanics and Hydraulics, University of Iowa, and a member of the Iowa Natural Resources Council, the state water resources and flood control agency.

doxical situation is, of course, that more value is continually being placed in the flood plain. This is expressed in terms of financial investment which is progressing on a large scale. We are all aware of the attractiveness of the flat level land so characteristic of rivers and flood plains and realize their historical role in the development of transportation and industry in this country. The flood plain has proven equally attractive for residential development and agriculture. Furthermore, because of its uncertain situation with regard to flooding, it is usually cheaper than land not subject to flood. Because of this and the ease with which construction can be carried out, flood plain land proves unusually attractive to people wishing to undertake enterprises which have low initial cost—low, that is, in comparison with the initial cost of alternative locations. When the inevitable flood occurs and great damage ensues, these people are the first to clamor for immediate emergency assistance which, as we have recently observed in Iowa, has to be provided at great expense to the governmental agencies involved. This phase is then followed by another in which financial support is requested from the federal government, or from the state or city, for flood protection works. The latter activity is a foregone conclusion after a great flood has subsided, as evidenced in several Iowa cities and towns, including Ottumwa, Sioux City, Waterloo, and most communities along the Mississippi River.[4,7] Although impressive and in large measure effective, such engineering works for flood protection as dams and reservoirs, levees, floodwalls, and channel improvements are costly measures; as a result, flood damages must be large to justify economically the cost of the proposed protection works. Thus the flood damage potential often remains unchanged and may even increase if investment continues unabated. Furthermore, complete flood control is not usually achieved; but the public, with complete faith in the protection works, rushes in with construction clear to the river bank, little realizing that damage frequency has been reduced, but not its inevitability.

This dilemma sets the stage for the subject matter of the chapters which follow, since obviously there must be a better, more rational way than the continued construction of engineering works for flood protection. The better way is the management of the flood plain, which to be effective must be brought about through political and legal means.[8] This requirement immediately imposes an impediment which is difficult to overcome. Often the owners of flood plain land will resist vehemently any attempt to regulate the use of their property. As an example, a policeman at Council Bluffs relates a story of an

# AN INTRODUCTORY PHILOSOPHY

old couple living in a cabin on the banks of Indian Creek. At the time of a flash flood on that stream, the police came to rescue the couple from the rapidly rising floodwaters. However, the old couple refused to leave, saying that they had lived in the cabin for many years and it had never been touched by a flood. Finally, the police carried them bodily from their cabin to safety. This action was taken none too soon, as flood waters soon completely inundated the area. This story was repeated more tragically at Sioux City in 1953. At least 14 persons drowned when flood warnings went unheeded, and the rapidly rising floodwaters of the Floyd River trapped many victims against ceilings, as depths exceeded the first-story level of the numerous bungalows in the low residential areas of the flood plain.[4]

Another type of property owner invariably opposing flood plain regulation or management is the one who wishes to place an apartment building, a motel, or other income-producing structure within the reach of floods. Sometimes he is merely subdividing and selling land on which other people may build houses; and if he carries on the sales activity in dry times, he may find ready victims. At Iowa City one subdivider saw his advertising sign "Choice Lots for Sale" submerged when flood releases from the Coralville Reservoir at half the expected maximum rate inundated the area. Such property owners are quick to claim violation of their constitutional rights and call any type of restrictive regulation "confiscatory" since it might deprive them of an income or increased profit from use or sale of their property. On the other hand, the public, who must become involved in the difficult tasks of rescuing people during floods, alleviating their discomfort with costly emergency measures, and trying to prevent the encroachment of floods through temporary and permanent measures, is in general unaware of this cost. Hence, the public is never as interested in promoting management as are profit-motivated people in preventing it.

Ignorance of the flood hazard also presents an impediment which can be surmounted only through technical knowledge accepted and acted upon by the political and legal representatives of the public. A most unfortunate incident which occurred at Muscatine, Iowa, in June, 1961, illustrates the need for wise and prudent public regulation. As the incident was related, a French woman with small children arrived destitute in Muscatine and appealed to the Salvation Army for shelter. These kind people found a cabin on the banks of Mad Creek and gave the family a home for the night. A great storm broke and a floodwave washed down Mad Creek, taking the cabin and its inhabitants with it. Obviously the woman could not have been expected to know of the flood hazard, nor did the Salvation Army have any idea of the danger. In fact, many people could be excused. However, all of us associated with flood plain management know the hazard which exists and have a moral obligation to prevent the occurrence of such accidents. Incidents such as this should convince all political and legal representatives that, although we might legitimately deny the construction of a house on a riverbank from the standpoint of reducing the conveyance of the flood plain, a more important facet exists, a moral obligation to prevent people from being drowned. In much the same fashion, legislation today prevents a manufacturer from indiscriminate marketing of dangerous drugs.

## FLOOD PLAIN MANAGEMENT AT IOWA CITY

The events experienced at Iowa City in the late 1950's and early 1960's serve as an excellent example of the problem facing flood plain managers. Enactment and enforcement of regulations, even though proper and reasonable, are not accomplished easily. To achieve even a measure of success requires long arduous hours of effort, a liberal amount of patience, and timely persuasion. Ultimately through such efforts a flood plain zoning ordinance was adopted at Iowa City. This was the first comprehensive flood plain ordinance in Iowa and as of 1967 remains the only one, although flood plain information studies are guiding subdivision regulations and urban development within existing ordinances at several localities. Less extensive, stopgap measures are in effect in several of these communities, coordinated with the flood plain regulatory activities of the Iowa Natural Resources Council (INRC), until more specific ordinances can be drafted and enacted.

The writer, by good fortune, was a member of the Iowa City Planning and Zoning Commission in the late 1950's, as well as a member of the INRC, and was thus able to participate actively and positively in securing the enactment of a flood plain zoning ordinance by Iowa City. Despite these favorable circumstances and the presence of a flood control reservoir only four miles above the city, several years actually were required to accomplish this.

Technical studies were made first to determine the flood hazard. The Corps of Engineers, Rock Island District, were consulted regarding operation of the Coralville flood control reservoir on the Iowa River.[6] With insufficient storage available to control maximum

# AN INTRODUCTORY PHILOSOPHY

experienced regional floods, high release rates must be expected during major flood periods. The residual flood hazard, including the flood potential of some 150 square miles of uncontrolled area lying between the reservoir and Iowa City, was evaluated in a study by the technical staff of the INRC.[5] Engineering and preliminary legal aspects of the problem were evaluated, and a flood plain identification and zoning method was presented for the city's use. This included a physical description and maps of the flood plain land which should be zoned, floodway encroachment limits within which an open use should be maintained to assure sufficient cross section for conveying flood discharges, the flood plain area which could be developed, and elevation controls to minimize damage in all areas.

Coordination between the state agency staff and the city officials at Iowa City was fostered by the INRC. Of particular value was the coordination effort of Clifford P. Peterson, an attorney and presently assistant director of the council, and Merwin D. Dougal, senior staff engineer at the time. The greatest task was reasoning with the city attorney, who was not convinced of the legality of any form of flood plain zoning. He approached the problem of controlling the use of such land with a great deal of skepticism. Although shown similar provisions contained in any zoning ordinance which regulated the property owner's use of his land, the attorney remained adamant until the flash flood at Muscatine occurred, as related previously. Following this tragedy, the need for regulation became apparent to him and he accepted its necessity.

A second setback occurred shortly thereafter when a new city attorney was appointed. After reviewing the zoning proposal, he agreed to continue only if the ominous words "inundation limit" and "encroachment limit" of the proposed flood plain zone boundaries were changed to "valley plain" and "valley channel" zones, respectively. Recognizing this would be more palatable to the real estate interests and local land subdividers, he then drafted a simple but adequate ordinance.[1] This was adopted as part of a new comprehensive zoning ordinance for the city on July 26, 1962.[2,3] Figure 1.1 shows a typical area controlled by the ordinance, with the valley districts outlined thereupon. Open uses are permitted in the valley channel zone (VC), no structures being permitted. Those structures allowed in the valley plain zone (VP) must conform to the permitted uses of land adjacent thereto as well as to the minimum elevations shown on the zoning map for the valley district.

Similar maps and data were provided to the Johnson County

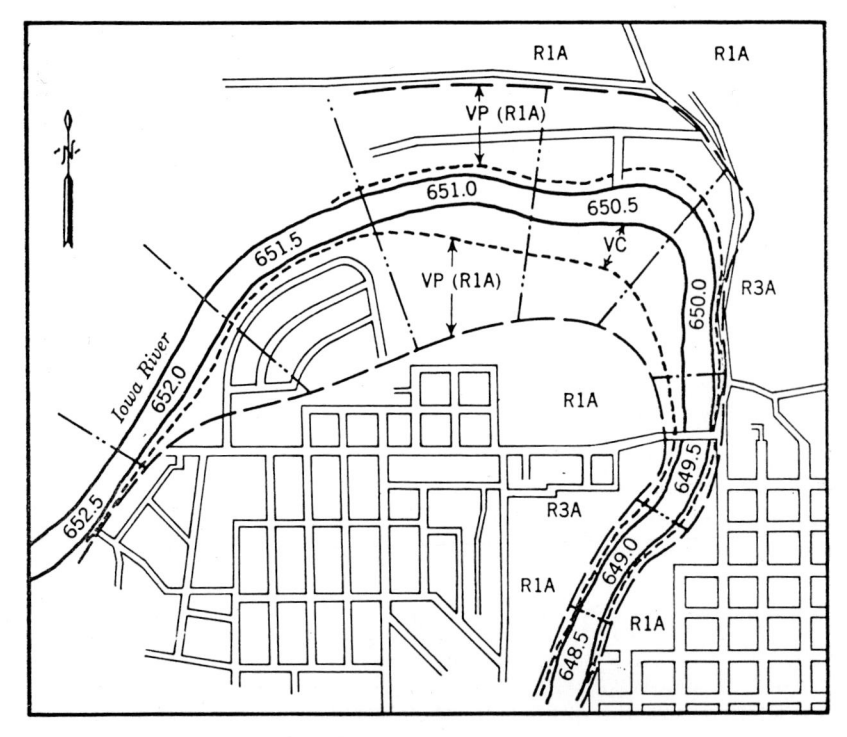

Fig. 1.1. A typical residential area controlled by flood plain zoning at Iowa City, Iowa (after Howe, 1963).

Zoning Commission and were used subsequently by the county in developing its regulation policies for suburban fringe developments. Construction of residences in the county along the river was progressing at a rate almost comparable to that within the corporate limits of the city.

Thus at Iowa City were encountered many of the problems and frustrations which often cause the delay of urgently needed programs for flood plain management, or which may entirely prevent their adoption.

## CONCLUSIONS

The optimum use of flood plain lands can be brought about only if those associated with their development and regulation gain the proper perspective and philosophy concerning the task lying before them. This brief introduction illustrates not only some of the objections commonly encountered but also provides an indication of the real need for flood plain management. Effective flood plain zoning ordinances can be enacted if those assigned the task will persevere in their efforts. The need for continued surveillance of the enforcement of enacted ordinances must also be recognized. These ideas should serve to illuminate the value of the chapters which follow.

## REFERENCES

1. Howe, J. W. Modern flood plain zoning ordinance adopted by Iowa City. *Civil Eng.* 33, no. 4 (Apr. 1963): 38–39.
2. Iowa City, City of. Zoning ordinance 2238. 1962.
3. Iowa City *Press-Citizen.* Aug. 7, 1962.
4. Iowa Natural Resources Council. *An inventory of water resources and water problems. For eight selected river basins in Iowa.* Bull. 1– Bull. 8. Des Moines. 1953–1959.
5. ———. *A study of flood problems and flood plain regulation, Iowa River and local tributaries at Iowa City, Iowa.* Mimeo. Des Moines. June 1960.
6. U.S. Army Corps of Engineers. *Coralville Reservoir, Iowa River, Iowa, regulation manual, preliminary (and amendments thereof).* Rock Island: U.S. Army Eng. Dist. 1951.
7. ———. *Interim review report for flood control at urban areas along the upper Mississippi River from Hampton, Illinois, to Mile 300.* Rock Island: U.S. Army Eng. Dist. 1961.
8. U.S. House of Representatives, Committee on Public Works. *A unified national program for managing flood losses.* Rept. of the Task Force on Federal Flood Control Policy, House Document 465, 89th Cong., 2nd sess. 1966.
9. White, G. F., et al. *Changes in urban occupancy of flood plains in the United States.* Dept. of Geog. Res. Paper 57. Chicago: Univ. of Chicago Press. 1958.

# 2

# MAN SHOULD MANAGE
# THE FLOOD PLAINS

### James E. Goddard

IT IS ENCOURAGING TODAY to observe the increased activity in flood plain management and the great interest in a subject that has only recently been given the major attention it deserves. It is also impressive to note the number of disciplines concerned—planners; economists; engineers; geographers; lawyers; naturalists; public officials; and staff members of local, state, university, and federal agencies. This professional interest and the combined effort of all these disciplines and agencies are necessary for effective management and wise use of one of our natural resources—the flood plains of our streams, lakes, and oceans.

This increased activity is in response to a nationwide attack on the unsolved problems of flood damages. In a national task force report on flood plain management policy recently transmitted to Congress, President Johnson declared, "To hold the nation's toll of

JAMES E. GODDARD is a private consultant in flood plain management at Tucson, Arizona, following his retirement as Chief, Local Flood Relations, Tennessee Valley Authority. He is a member of the national Task Force on Federal Flood Control Policy.

flood losses in check and to promote wise use of its valley lands requires new and imaginative action." He concluded in his transmittal letter:

> There is a role for each level of government in a successful flood damage abatement program. There is likewise a responsibility on all participants, from the individual citizen through many elements of Federal establishment, to contribute to the program's success. Let us begin today a renewed and cooperative effort to attack this problem.[11]

As a result of this new and increased emphasis, flood plain management is rapidly becoming a challenging and rewarding field in the total spectrum of water resources. The purpose of this discussion is to reflect upon its real meaning, review the historic role of flood plain utilization and attendant flood damages, and enlarge upon the need for comprehensive and broad approaches to management. Recommendations for more effective action are included in the final section.

## THE MEANING OF FLOOD PLAIN MANAGEMENT

It is most appropriate to begin by clarifying the scope and definition of "flood plain management." This term includes all measures for planning and action which are needed to determine, implement, revise, and update comprehensive plans for the wise use of flood plain lands and their related water resources for the welfare of our nation.

The term "flood plain management" was coined a few years ago as a part of coordinated efforts of federal, state, and local officials and competent water resources engineers, geographers, planners, economists, lawyers, foresters, recreation specialists, naturalists, and others concerning an overall comprehensive approach to flood plain problems. Each group naturally preferred a term that more nearly expressed and would tend to give some emphasis to its particular philosophy. For example, many engineers continue to claim that the term "flood control" is all-inclusive in its meaning, and many planners think of "management" as regulation of land use. Consideration had to be given not only to the technical definitions of terms but, more important, to the public interpretation and understanding of selected terminology. The term flood plain management appeared to be the most acceptable (or the least objectionable) for the overall approach. It has now been widely accepted and is being used more and more by the several agencies and many officials concerned with flood plain activities.[6,9,11]

The term includes flood control (single-purpose or multipurpose projects) and all other alternative actions. Some professional people still refer to "flood control *and* flood plain management" either through misunderstanding or because they are thinking of management only as a regulation of land use and have difficulty in withdrawing from the belief that flood control is a cure-all. This is as unfortunate as the use of such terms as "flood plain zoning" instead of flood plain regulations; zoning is only one of the tools in regulation, both being encompassed within the field of management.

The purpose of a national flood plain management policy can be stated in terms of two broad concepts.[3] First, we must minimize the loss of life, personal suffering, and physical hardships which are the immediate consequences of severe floods. Second, we must endeavor to achieve in the long run the optimum economical use of the nation's flood plains, with due regard for both private and public benefits and related costs. As similarly expressed in the report of the Bureau of the Budget's Task Force on Federal Flood Control Policy, for a unified national program for managing flood losses, public policy must distinguish "between the problems of minimizing damage to existing flood plain developments and the problem of achieving optimum future use of flood plains."[11] In accomplishing this twofold purpose, we will be (1) evaluating through physical and economic concepts the desirability of protecting investments already made; and (2) of equal or greater importance, concerning ourselves with selecting in flood plain areas the best investment alternative from a combination of the many development possibilities now available.

## FLOOD PLAINS AND FLOOD DAMAGES

Flood plains have played an important role in the development of civilization and of our nation, both in terms of transportation and water use and in availability of supply.[14,15] The major role of past years has gradually diminished, as technological advances have altered the relative advantages of flood plain locations. Nevertheless, the economic advantages of using the small percentage of flood plain lands, especially those along selected coasts and in major delta areas, must be carefully considered to ensure the optimum growth of the social and economic welfare of these regions.

The growth of the flood damage potential through the unwise use of flood plains has been recognized and documented in recent years.[11,17] In brief, despite total expenditures exceeding $7 billion and continuing annual expenditures of more than $500 million for efficiently designed and constructed flood control structures, the annual losses and suffering from floods are gradually mounting.

Construction of flood control structures has been unable to keep pace because of a lack of positive control over events causing an increase in the damage potential and actual losses. Flood plain construction continues (1) in or near older areas where flood protection works never were feasible; (2) in flood-prone lands adjoining but outside protected areas; (3) in protected areas where damage and extraordinary losses are or will be experienced when catastrophic floods exceed the design flood criteria; and (4) in rapidly urbanizing areas where previous flooding was never well documented, and the flood hazard may not easily be recognized or may be ignored.[11] Because flood control through protection works cannot physically or economically eliminate the damage potential, flood damages will remain a problem as long as flood plains continue to be occupied by man.

## THE ROLE OF SUBSIDIES

The role of subsidies in our social structure is worthy of review. It is difficult to say when or how the use of subsidies started. Perhaps one of the first was the dowry of a bride! But a few of the more prominent in the minds of the present generation would be those granted to railroads, airlines, housing, urban renewal, model cities, industry for national defense, agricultural producers, and flood plain occupants. Some subsidies are directed toward physical problems, others toward social problems relating to human behavior, and many toward a combination of the two. The relative importance of various subsidies is continually changing in our dynamic environment to reflect current conditions and needs of the nation.

The flood plain problem was originally a combination of the physical and the social.[11,14] The need for water, transportation, power, agricultural products, and other items critical to the development of a region justified certain subsidies which encouraged or protected river and flood plain development. But technical advancements have tended to relegate the problem to a physical one, with exceptions that include major coastal and delta areas where occupancy cannot be avoided. To what degree is it wise to continue subsidizing this problem? Should we begin to subsidize the leveling of hills to reduce erosion and other attendant development costs—another physical problem? Or take other similar actions? It is well to remember there is a limit on public funds available for subsidies and public welfare;

therefore, allocated funds should be used where they will serve our nation best.

Continued subsidies in the area of flood plain occupancy will further encourage a greater use than would be justified by the strict application of the economics of alternative locations. Our three levels of government—local, state, and federal—are assuming ever larger obligations to remedy flood losses; to coordinate emergency efforts during floods; to make provisions for emergency relief; and to offer aid for cleanup, repairs, and reconstruction. The collective nature of the benefits from this public subsidy is at the expense of all the taxpayers. A time for change is at hand, with the promise of making those who occupy the flood plain more responsible for the results of their actions.

The recent national task force introduced a new, far-reaching concept which might in the future provide a more effective solution to indiscriminate occupancy.[11] This concept involves both flood plain occupancy charges and indemnification of flood losses. Mandatory, actuarial, risk-related, annual occupancy charges would be imposed. The annual payment would be equal both to the occupant's estimated annual damages and to any external costs which his occupancy causes to other individuals or to the public. These payments would be placed in a collective indemnification fund from which compensation would be made to those experiencing flood losses.

In this manner the full costs of flood plain occupancy would be placed on the actual or prospective occupants. The proposed concept provides for a means of budgeting flood losses from which there is no complete escape, even with flood protection works. New developments would then be limited to those from which a real economic advantage could be anticipated. The total net social income from flood plain use, including both private and public benefits and costs, would be at an optimum if the proposed system were accepted and implemented.

However, such a comprehensive program is premature for the present. We do not have the risk element sufficiently established, nor are the administrative means available. Other management methods must be utilized until this more comprehensive concept can be introduced.

## ACCEPTANCE OF LAND-USE PLANNING AND REGULATION

Flood control in its many forms started long ago.[14] But land-use planning as such began in the 1920's. The need for regulating land use for public welfare became more apparent as cities grew, congestion increased, and industrial growth accelerated. Regulations were increasingly accepted by the public and the courts. But only since the 1950's has there been an increasing realization that the use of flood plains must be wisely planned and regulated for the benefit of the public's health, safety, and welfare.[2,5,11]

Planning, zoning, and other guides for use of lands have become more and more effective as our nation continues its seemingly boundless growth. Emergence from the agricultural nation of only three decades ago into an urban industrial society has increased pressures to move into flood hazard areas.[16,17] It has also brought about other related changes in water resources that are presenting serious challenges but which lead to improved procedures in water resources management. Water supply and water quality control are examples of other problems caused by intensified urban and rural land use.

There has been a gradual broadening of the approach to managing the flood plains. One of the most important and effective elements has been the increasing degree of cooperation between local, state, and federal agencies. Many of the problems are too large for local or even state government to solve alone. It is neither feasible nor desirable for the federal government to act independently. We must recognize that authority and responsibility for planning, zoning, and controlling local land use lies solely with the states and their local government units. However, the magnitude of the problem, national in scope, demands federal attention and participation. Through collective action a real measure of success can be achieved.

## THE GROWTH OF PUBLIC AWARENESS

The increasing awareness of the need for improved management and wise judgment in use of our flood plain resources is evidenced by the many studies, reports, and actions which have been published. A bibliography of these is printed periodically by the Tennessee Valley Authority (TVA).[8] The conclusions and programs recommended in all of them acknowledge the need for a broader approach to water resources development, flood damage reduction, and flood plain management.

The TVA, an agency of the federal government, pioneered a program of flood damage prevention starting in 1953.[4] In 1959 it submitted to the Congress recommendations for a national program of flood damage reduction. A review of the national scene in 1959 by

# MAN SHOULD MANAGE THE FLOOD PLAINS

the U.S. Senate Select Committee on National Water Resources included the many problems resulting from man's encroachment on flood plains.[13] Among its recommendations was the consideration of all alternatives in determining the most economic and acceptable solution of flood problems. The Department of the Army Civil Works Study Board made a broad survey of its civil works program in 1964 and the recommendations are now leading to greater emphasis on alternatives and to closer liaison between federal and state water resources planning agencies. The federal Water Resources Planning Act of 1965 encourages and stresses consideration of all reasonable alternatives and cooperation of all federal agencies, states, local governments, individuals, corporations, and others concerned.[10]

In 1965 the Office of Emergency Planning arranged for a review of the flood problem and made recommendations concerning federal policy in grant, loan, and construction programs for areas with high risk of natural disaster damage. At the request of Congress, the Department of Housing and Urban Development studied flood insurance and indemnification against natural disasters and submitted its recommendations in 1966.[12] The Bureau of the Budget's Task Force on Federal Flood Control Policy made a comprehensive review of the nationwide flood situation in 1965–1966.[11] Its report calls for separate emphasis by various levels of government and property owners, to be achieved through greater coordination of federal efforts and increased roles by the states. Executive directives and congressional authority and appropriations have initiated the broad-concept program recommended by that task force.[6,9,11]

Many states have recognized the need for considering alternative approaches to their flood problems and for encouraging the wise use of their flood plains. This recognition invariably has occurred as the lack of funds and other factors have delayed flood control projects and problems became increasingly acute. Several states adopted encroachment laws of some type years ago, but some of them have not always used the authority effectively. Others have more recently adopted legislation and programs that go far toward sound flood plain management. Among these are California, Connecticut, Iowa, Nebraska, New Hampshire, New Jersey, North Carolina, Tennessee, and Wisconsin. Others have legislation pending.

A number of cities and counties have studied their flood plain problems and have initiated management programs, and others are now in the process of review. Cities such as Boulder, Colo.; Bristol, Tenn.-Va.; Detroit, Mich.; Hartford, Conn.; and Knoxville, Tenn., have made substantial progress. The northeast Illinois Metropolitan Area Planning Commission made an extensive study for the Chicago metropolitan area.

Statements, conferences, and committee actions by the Council of State Governments and interstate compacts such as the Delaware River Basin Commission encourage wise decisions and establishment of standards for flood plain use. Various studies have been made by university research groups, most notably the one led by Gilbert F. White at the University of Chicago.[5,7,16-18] Professional societies such as the American Institute of Planners and the American Society of Civil Engineers have issued reports and guides.[1,2]

## BROADER PROGRAMS NOW BEING DEVELOPED

Currently, several federal and state programs are actively giving broader consideration of alternatives and assisting in the wise management of flood plain resources. These are growing in number and effectiveness. The TVA continues its local flood relations program. In addition to the nearly three-score communities which have adopted flood plain regulations, several have been assisted in planning for comprehensive flood damage prevention and in the necessary coordination and other local action needed to put those plans into effect.

The Corps of Engineers launched a nationwide flood plain management services program a little more than a year ago.[9] The director of civil works of the Corps of Engineers and his technical staff met with the nearly 50 division and district engineers and selected members of their staffs in a series of planning policy conferences during November, 1967. Basic goals in broad resource planning were reviewed, expanded consideration of alternatives was discussed and underscored, external coordination was emphasized, and benefit evaluation problems were explored. It is hoped these conferences will lead to more applicatory and imaginative planning vital to the proper development of the nation's flood plains and related water resources.

President Johnson's Executive Order 11296 requires all federal executive agencies to evaluate flood hazards in locating federal facilities, in disposing of federal lands and properties, and in administering federal grants and loan or mortgage programs.[6] The U.S. Geological Survey provides flood information through its hydrologic atlas program, as reported in another chapter of this book. The Department of Housing and Urban Development will provide guidance for planning the use of flood plain lands, and is cooperating with Congress concerning the federal flood insurance legislation. Federal legislation has been approved by the House and the Senate, but in somewhat different forms. Other federal agencies provide information related to their programs, including the Environmental Science Services Administration (Weather Bureau), Soil Conservation Service, Bureau of Reclamation, and Bureau of Outdoor Recreation.

California is administering its Cobey-Alquist Flood Plain Management Act. Connecticut continues to establish encroachment lines delineating open floodways for its streams, and the state also coordinated with the City of Hartford and federal agencies in a comprehensive flood plain management project. Iowa's coordinated method of guidance and assistance to communities in studying, planning, and acting on their local flood problems is commendable. New Jersey's program includes preparation of flood plain information reports to supplement federal efforts. Tennessee continues its effective program of assisting communities in a strong statewide program and is assisting communities with regulations and wise management of flood plains. Nebraska and Wisconsin have recently initiated flood plain management programs. Other states are also active in this area.

Cities and communities are also taking positive and effective actions. Coordinated planning by local, state, and federal governments is leading to an increasing number of local comprehensive plans in which single-purpose and multipurpose structures are combined with floodproofing, regulations, and other alternatives for the most effective and acceptable projects. These projects are introducing innovations in public understanding, cooperative relationships, evaluation of benefits, costsharing, financing, and other factors. Several of these forward-looking municipalities were mentioned previously. The flood plain management program for the Chicago metropolitan area is an example of comprehensive action.

## SUCCESS THROUGH COOPERATION

Unilateral decisions and actions can no longer provide acceptable answers to our problems. Each level of government and each individual have a role. The federal interest is unquestioned. State and local governments and individual owners of properties in flood plains also must face up to their responsibility if there is to be improved management.

Neither is the task one for engineers alone, or economists alone, or planners alone, or any one discipline alone. The magnitude of the problem and the myriad possible solutions or combinations to be considered demand contributions from many disciplines.

## RECOMMENDATIONS FOR MORE EFFECTIVE ACTION

Some suggestions and recommendations are offered for expediting and increasing effective action concerning problems that are daily becoming more critical in a growing number of areas. Perhaps these will lead to intensified timely assistance that will result in a reduction both in human suffering and in dollar damage.

1. Improve public relations and develop positive programs for informing the public, so that, with general understanding of the need, there will be necessary public support for action programs.
2. Arrange conferences in each state similar to the Iowa conference of 1968 for representatives of all disciplines and levels of government. These should be directed toward developing better understanding of the respective federal, state, and local roles and of the relationship between disciplines in flood plain management.
3. Accelerate the effort of the federal government by (a) outlining the broad concepts and all possible alternatives which should be given greater consideration in planning the proper utilization of flood plain lands and of related water resources activities; (b) increasing the annual production of flood plain information reports, so that basic information will be available within 15 years for the 7,500 cities and communities with flood problems; (c) providing greater assistance and guidance to the states and local governments in design of flood plain regulations and use of flood data in decisions concerning use of individual sites in flood plains; and (d) giving further consideration to flood insurance, and other forms of indemnification against natural disasters, in the form of pilot programs and other research.
4. Enlarge the roles of the states and encourage them to accept greater responsibility in programs to assist their citizens through (a) legislative and executive recognition of the need for greater action; (b) legislation to authorize (and executive action to establish) new agencies or effectively reorganize existing state agencies, to prepare statewide water resource plans including flood plain management, to establish flood plain regulations either by state or local authority, to provide guidance and assistance to local governments, and for related actions; and (c) increasing the contribution, operation-

ally or financially, for the preparation of flood plain information reports to ensure their availability for communities before existing conditions become too critical and before more optimum economic benefits are jeopardized by unwise use of flood plains.
5. Increase the role of local governments (city, county, and regional) and their acceptance of greater responsibility through (a) adoption of flood plain regulations, (b) provision of guidance and technical assistance for floodproofing of buildings, and (c) provision for increased interest and participation in planning and executing comprehensive flood plain management and flood damage prevention projects.

In addition, we must improve attitudes and cooperation between those professions and decision makers most involved in flood plain management problems and assist them in achieving workable solutions. Only close relationships at the working level between the elected and appointed officials, the engineers, the planners, the economists, the scientists, those in the academic field, the geographers, the business administrators, the lawyers, the naturalists, and others, coupled with a mutual desire to best serve man's rapidly changing society, can accomplish this. Flood plain management will then become a reality.

## REFERENCES

1. American Society of Civil Engineers. Guide for the development of flood plain regulations. Progress rept., Task Force on Flood Plain Regulations. *Proc. Am. Soc. Civil Eng., Hydraulics Div.* 88, no. HY5, Paper 3264 (Sept. 1962).
2. American Society of Planning Officials. *Flood plain regulation.* Planning Advisory Serv. Rept. 53. Aug. 1953.
3. Caulfield, H. R., Jr. Flood plain management policies. *Proc., National Conference of State and Federal Water Officials.* Washington: Water Resources Council. 1967.
4. Goddard, J. E. The cooperative program in the Tennessee Valley. In G. F. White, ed. *Papers on flood problems.* Dept. of Geog. Res. Paper 70. Chicago: Univ. of Chicago Press. 1961.
5. Murphy, F. C. *Regulating flood plain development.* Dept. of Geog. Res. Paper 56. Chicago: Univ. of Chicago Press. 1958.
6. Office of the President of the United States. *Evaluation of flood hazard in locating federally owned or financed buildings, roads, and other facilities and in disposing of federal lands and properties.* Exec. Order 11296. 1966.
7. Sheaffer, J. R. *Flood proofing: An element in a flood damage reduction program.* Dept. of Geog. Res. Paper 65. Chicago: Univ. of Chicago Press. 1960.

8. Tennessee Valley Authority. *Flood damage prevention, an indexed bibliography,* 5th edit. Knoxville. 1967.
9. U.S. Army Corps of Engineers. *Water resources development, functions, and programs of the Corps of Engineers.* Washington: Chief of Engineers. May 1967.
10. U.S. Congress. *Water resources planning act of 1965.* PL 89-90. 89th Cong., 1st sess. 1965.
11. U.S. House of Representatives, Committee on Public Works. *A unified national program for managing flood losses.* Rept. of the Task Force on Federal Flood Control Policy. House Document 465, 89th Cong., 2nd sess. 1966.
12. U.S. Senate, Committee on Banking and Currency. *Insurance and other programs for financial assistance to flood victims.* 89th Cong., 2nd sess. 1966.
13. U.S. Senate, Select Committee on National Water Resources. *Flood problems and management in the Tennessee River basin.* Committee Print 16, 86th Cong., 1st sess. 1959.
14. ———. *Floods and flood control.* Committee Print 15, 86th Cong., 2nd sess. 1960.
15. ———. *Future needs for navigation.* Committee Print 11, 86th Cong., 2nd sess. 1960.
16. White, G. F. *Human adjustment to floods: A geographical approach to the flood problem in the United States.* Dept. of Geog. Res. Paper 29. Chicago: Univ. of Chicago Press. 1942.
17. White, G. F., et al. *Changes in urban occupancy of flood plains in the United States.* Dept. of Geog. Res. Paper 57. Chicago: Univ. of Chicago Press. 1958.
18. White, G. F., ed. *Papers on flood problems.* Dept. of Geog. Res. Paper 70. Chicago: Univ. of Chicago Press. 1961.

Mississippi River flood at Dubuque, 1965. Courtesy Corps of Engineers, Rock Island District.

# Part 2

*There is no reason to assume that the flood of record will not be exceeded by a yet greater disaster.*

Wolf Roder

# FLOODS AND THE FLOOD POTENTIAL

# 3

# FLOODS IN IOWA

### Harlan H. Schwob

GREAT FLOODS have plagued mankind from ancient times. The flood that introduced Noah to the ship-building and animal husbandry trades is perhaps the first to be recorded. However, floods undoubtedly occurred prior to this event and certainly have occurred since. In Iowa we have only a short record of these events, much less than a hundred years, and it probably is not all-inclusive or representative of a very long period, such as a thousand years or more. Nevertheless, this record contains some data of interest on outstanding floods. It is my purpose to describe briefly the magnitude of the greatest floods of record and the storms which caused them. This will provide a reference level for the flood plain management topics which follow. Data obtained and filed by the U.S. Geological Survey were used in making this study, with much of the basic data having been previously published in the annual bulletins.[7,8] Several special studies were a source of additional information, as will be noted.

HARLAN H. SCHWOB is Hydraulic Engineer in charge of the special studies section of the U.S. Geological Survey, Iowa City, and the author of several bulletins concerning Iowa flood discharge and frequency experience and stream low-flow characteristics. Publication authorized by the Director, U.S. Geological Survey.

## COMPARING PEAK DISCHARGES

The maximum floods of record can be described and compared in a quantitative sense only if a frame of reference is established from which the outstanding floods can be ascertained. The technical and statistical framework within which an analysis can be made of flood peak discharges in Iowa was presented first in a 1953 bulletin on Iowa floods.[3] This early study was updated in 1966 in a study sponsored by the Iowa Highway Research Board (Bulletin No. 28), using the additional years of hydrologic record and more advanced statistical and electronic computer methods.[4] This latter report, in which the magnitude and frequency of Iowa floods is most recently reported, will serve as the frame of reference for comparative purposes and as a source of the flood data.

Multiple regression analysis was made of the flood data for Iowa rivers. This resulted in the division of the state into two hydrologic regions, with a formula being developed for each region permitting the mean annual flood (MAF) to be computed. The variables include drainage area, stream slope, and for most of the state the normal annual precipitation. The mean annual flood is described quantitatively as the mean of the annual flood peak discharges that would be derived from a very long streamflow record. At many places in the state this flood is approximately a bankfull stage. The formulas are not essential to the purpose of this chapter, but it should be noted that they permit the computation of the mean annual flood on nearly all streams in the state where adequate topographic maps are available. Using the mean annual flood as a basis for quantitative comparison, Figure 3.1 shows the relation of major peak discharges to the mean annual flood discharge.

The magnitudes of several great floods recorded at stream-gaging stations or at miscellaneous measurement sites have been plotted in Figure 3.1 using separate symbols for each type of station. Also shown in Figure 3.1 are several lines representing ratios to the mean annual flood. The uppermost dashed line with a ratio of 40 times the mean annual flood discharge encompasses all the floods. Two other dashed lines indicate ratios of 20 and 10 times the mean annual flood. The two solid lines are the 50-year recurrence-interval floods for the two flood frequency regions of the state as shown in Bulletin No. 28. They have a ratio of 4.50 for a small area in northwest Iowa and 3.10 times the mean annual flood for the remainder of the state.

The 25 floods that are above the 10-unit ratio line were caused

# FLOODS IN IOWA

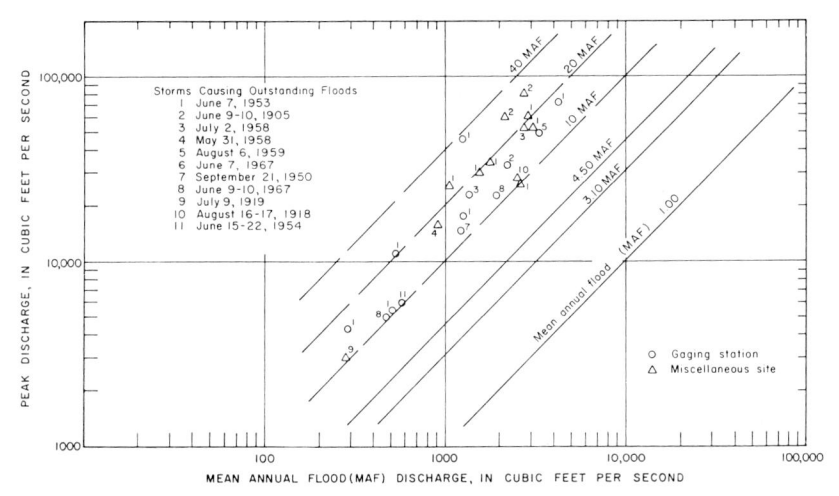

Fig. 3.1. Relation of peak discharge to mean annual flood discharge for outstanding floods within Iowa.

by 11 storms. There are 6 floods with ratios between 20 and 40 that resulted from 2 storms. The 11 storms are listed in Figure 3.1, approximately in descending order of the magnitude of the peak discharges they produced. The reason for a larger number of discharges than storms is that peak discharges were obtained at more than one place in a basin or in two or more nearby basins for any single storm.

## ELEVEN MAJOR IOWA STORMS

Each of the eleven storms that caused outstanding floods will be described briefly. The magnitudes of the resulting floods will be indicated in terms of ratio to mean annual flood. The general location of the storms is shown in Figure 3.2. Each circle and number indicates the location of a storm, but does not indicate its extent or intensity. Storms numbered 4, 6, 9, and 10 are of limited extent and duration but of very high rainfall intensity. They illustrate the cloudburst activity frequently experienced during severe thunderstorms. The others covered a much larger area, also at a relatively high rainfall intensity but of somewhat longer duration.

The first of the storms to be described produced outstanding floods in northwestern Iowa on June 7, 1953.[11] This storm had two centers, one over the Floyd River basin and the second over the upper reaches of the Little Sioux River. In this area total rainfall

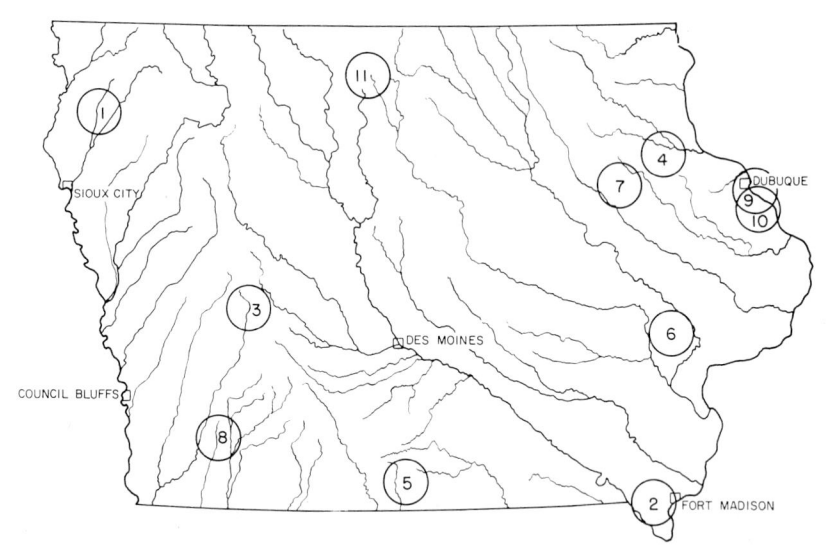

Fig. 3.2. Outline map of Iowa showing approximate location of centers of outstanding storms in Iowa.

amounts of 10.75 and 11 inches in about 14 hours were reported at two locations. Rainfall in excess of 7 inches was general over a large part of the northwest quarter of the state. Damages and loss of life were high, particularly in Sioux City and in the remainder of the Floyd River basin. In Figure 3.1, four of the six peak discharges plotting between 20 and 40 times the mean annual flood were produced by this storm. Additionally, 8 of the 19 peaks that lie between 10 and 20 times the mean annual flood also resulted.

The second storm, on June 9 and 10, 1905, occurred in southeastern Iowa and adjoining areas of Missouri and Illinois.[1] Maximum rainfall in Iowa apparently occurred in the vicinity of Bonaparte (in Van Buren County about 27 miles west and slightly north of Fort Madison), with a recorded total of 12.10 inches in 12 hours. A second center of heavy rainfall was at La Harpe, Illinois (about 20 miles east of Fort Madison), where 10.25 inches in 12 hours were recorded. The runoff from this storm produced floods of great size on many streams in the area, particularly in the Devil Creek basin located immediately west of Fort Madison. Rainfall over this basin varied from about 7 inches near the mouth to over 10 inches near the northwest corner. Two estimates of the peak runoff have ratios of 28 and 29 times the mean annual flood from drainage areas of 109

# FLOODS IN IOWA

and 152 square miles. Flood discharge determinations from this storm account for two of the floods in Figure 3.1 in the 20 to 40 range of ratios, and one additional determination in the 10 to 20 range of ratios.

The third storm, on July 2, 1958, struck the headwater areas of the Nishnabotna and Raccoon River basins and was nearly centered over the city of Audubon in the west-central part of the state, where the rainfall was estimated as exceeding 13 inches in 24 hours.[9] Flash floods in the headwater tributaries caused 17 deaths and severe property damage in urban and rural areas of Audubon and Guthrie counties.[6] Two of the 19 flood peaks in Figure 3.1 that plot below the 20-unit ratio line were the result of this storm. Drainage areas at the two sites were 26 and 111 square miles. If this storm had been centered over a single basin rather than on the divide between the two basins, much greater peak discharges undoubtedly would have occurred.

The fourth storm that produced an outstanding flood occurred on May 31, 1958. This storm was small in areal extent but apparently was of great intensity. It was centered over the Wayman Creek basin in northeast Iowa near the town of Garber in Clayton County. No official Weather Bureau precipitation gages were in or near the area of heaviest rainfall. Unofficially, it was estimated that 6 inches of rain fell in 3 hours over the 7-square-mile basin. The result was a peak discharge 17 times the mean annual flood.

The fifth storm was that of August 6, 1959, over the south-central part of the state.[10] Rainfall of about 17 inches in 9 hours was reported at two places in Decatur County. A stream-gaging station on the Weldon River near Leon experienced a peak discharge 14.5 times the mean annual flood.

The sixth storm, on June 7, 1967, was over the Wapsinonoc Creek basin in Cedar and Muscatine counties.[2] Rainfall ranging from 4 to 13 inches in 14 hours was nearly centered on the 180-square-mile basin. At points on both branches of the creek near West Liberty the peak discharge was 12.5 and 13.7 times the mean annual flood from drainage areas of about 46 square miles each.

The seventh storm occurred on September 21, 1950, in northeast Iowa between Independence and Dubuque. Rainfall of 6.35 inches in 24 hours at Independence was probably exceeded at Winthrop, where the discharge of Pine Creek was measured. The peak discharge from 28.3 square miles was 12 times the mean annual flood.

The eighth storm, or series of storms, was that of June 9–10, 1967, over the southwest part of the state. Scattered rain of about 6 inches on sodden ground caused flash flooding on many small streams. The stream-gaging station on the Tarkio River at Stanton (about 8 miles east of Red Oak) had a peak discharge of 12 times the mean annual flood from 49.3 square miles. A tributary of the river had a peak discharge of slightly over 10 times the mean annual flood from a drainage area of 4.66 square miles.

The ninth storm was on July 9, 1919, in the Dubuque area. The Dubuque precipitation gage recorded a total of 3.87 inches of rain in 5 hours, with 2.23 inches falling in a period of 1 hour near the end of the storm. Union Park Creek, located near the northwest part of the city, had a peak estimated at about 11 times the mean annual flood from a drainage area of about 1 square mile.

The tenth storm occurred on August 16–17, 1918, in northeast Iowa. The Weather Bureau precipitation gage at Dubuque recorded a total 24-hour rainfall of 5.22 inches. Of this amount 2.96 inches fell in a period of 2 hours. Catfish Creek, located just south of Dubuque, produced an estimated discharge 11 times greater than the mean annual flood from a drainage area of 40 square miles.

The eleventh storm occurred over large areas in north-central and western Iowa on June 15–22, 1954.[12] The terrain is flat to gently rolling in much of this area. Total rainfall for the period ranged from 4 to 12 inches. One stream, draining 133 square miles in the upper end of the Iowa River basin, produced a peak discharge of slightly over 10 times the mean annual flood.

### ENVELOPE CURVES FOR THE GREATEST FLOODS

We also need to examine the relation of the magnitude of the flood discharge to the drainage area that produced it. Because the variables of stream slope and precipitation also affect the computation of the mean annual flood, the immediate effect of the size of the drainage area is not obvious in Figure 3.1; however, Figure 3.3 clearly shows the relation of the drainage area. In this figure the maximum floods in terms of ratio to the mean annual flood have been plotted against drainage area. The two straight-line segments are enveloping curves of the greatest floods for which we have data. The peak in the curve occurs at 265 square miles and a ratio of 36:1 and is defined by a flood that resulted from the first storm reported

# FLOODS IN IOWA

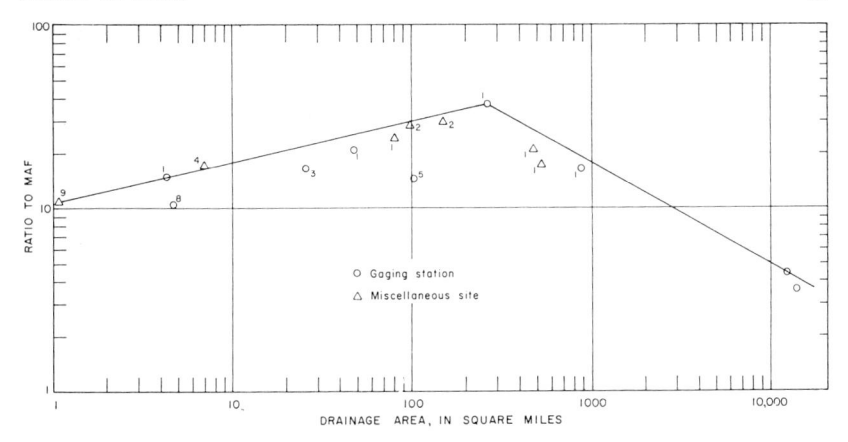

Fig. 3.3. Relation of ratios to mean annual flood and drainage area for outstanding flood peaks within Iowa.

previously—in northwest Iowa. At the 1-square-mile drainage area the flood produced by the ninth storm (at Dubuque) defines the location of the line at a ratio of 10.1:1. At 12,500 square miles and a ratio of 4.24:1 the location of the line is defined by the flood that occurred on June 14, 1947, on the Des Moines River near Tracy. The flood ceiling, as indicated by the curve, envelopes all known floods but will undoubtedly change as or if greater storms occur. The present record of floods is comparatively short, even at gaging stations on large streams. For locations with drainage areas less than 300 square miles, both the flood and storm records are a very small sample in time and number of sites.

Several additional flood characteristics should be noted. Inspection of the list of storms which caused the eleven major floods indicates that they were all in spring and summer. Furthermore, over half of these events occurred in June if we include the storm of May 31, 1958, in the June classification. Hydrologic records indicate that nearly all major streams in Iowa have experienced their peak discharge of record during the summer months. Snowmelt or a combination of snowmelt and rain have produced peak discharges of record at a few places in Iowa; however, their flood ratios were not large enough to be included in either Figure 3.1 or 3.3. The Mississippi River, however, is an exception. The flood in 1965, caused by a combination of snowmelt and rain, produced peak discharges of record at all points on the main stem upstream of the confluence with the Skunk River.[5] At McGregor the peak was about 2.6 times the mean annual flood. This was the maximum ratio experienced on the Mississippi River along the eastern Iowa border.

## CONCLUSIONS

The use of the mean annual flood to help define outstanding flood discharges is an attempt to place the data for different areas of the state on a common basis. It may appear that this attempt fails; for instance, the flat north-central area of the state has produced only one known flood exceeding a ratio of 10 times the mean annual flood. Although great storms are known to have occurred in this area of relatively flat terrain, they simply have not produced peak discharges comparable to those occurring in the rougher areas of the state. Present data seem to indicate that the flat areas are incapable of producing the high-ratio peak discharges characteristic of the floods most destructive to life and property. One possible reason for this is the large amount of surface, pothole, and valley storage available. However, these areas may experience relatively long periods of shallow inundation, with attendant flood damages. Inspection of Figure 3.3 indicates also the possibility that very large and very small drainage areas anywhere in the state may lack the flood ratio potential of the intermediate-size areas.

In conclusion, it must again be emphasized that the flood data presented are historical in nature. They serve to point out the potential and the extent of major floods known to have occurred within Iowa. These data will undoubtedly assist in planning programs for flood plain management. However, these results need to be supplemented by other studies, particularly those of an economic nature, in arriving at a solution to individual flood problems. Also, it should be noted that the selected ratio of peak discharge to mean annual flood of 10 as a lower limit was arbitrarily selected to reduce the amount of data and time required for a brief study and report. Many floods with smaller ratios have occurred, and a number have earned the classification of being a disaster, especially in the larger drainage basins. Furthermore, even floods with ratios of about 3 are considered to be relatively rare events. They would have a probability of about 2 percent of occurring in any given year at any point within a large area of the state.

## REFERENCES

1. Mead, D. W. *Hydrology*. Ch. 11. New York: McGraw-Hill. 1919.
2. Schwob, H. H. *Flood of June 7, 1967, in the Wapsinonoc Creek basin, Iowa*. USGS open-file rept. Mimeo. Iowa City. 1968.
3. ———. *Iowa floods, magnitude and frequency*. Iowa Highway Res. Board Bull. 1. Ames. 1953.
4. ———. *Magnitude and frequency of Iowa floods*. Iowa Highway Res. Board Bull. 28, pt. 1. Ames. 1966.
5. Schwob, H. H., and Meyers, R. E. *The 1965 Mississippi River flood in Iowa*. USGS open-file rept. Mimeo. Iowa City. 1965.
6. U.S. Army Corps of Engineers. *Report on the flood of July, 1958, in the Nishnabotna River basin, Iowa*. Mimeo. Omaha: U.S. Army Eng. Dist. 1959.
7. U.S. Geological Survey. *Water resources data for Iowa*. Surface water records of Iowa, pt. 1. Iowa City. Publ. annually, 1961 to date.
8. ———. Water supply papers. Washington: USGPO. Annual, through 1960. *Part 5. Hudson Bay and Upper Mississippi River basins; Part 6-A. Missouri River basin above Sioux City, Iowa;* and *Part 6-B. Missouri River basin below Sioux City, Iowa*.
9. U.S. Weather Bureau. *Climatological data, Iowa*. Washington: USGPO. July 1958.
10. ———. *Climatological data, Iowa*. Washington: USGPO. Aug. 1959.
11. Wells, J. V. B. *Floods of June, 1953, in northwestern Iowa*. USGS WSP 1320-A. Washington: USGPO. 1955.
12. Yost, I. D. *Floods of June, 1954, in Iowa*. USGS WSP 1370-A. Washington: USGPO. 1958.

# 4

# THE FLOOD POTENTIAL
# AND FUTURE FLOOD PROBLEMS

J. Wm. Funk

THE HISTORIC FLOOD RECORD, as reviewed in terms of (1) severe storm phenomena, (2) the outstanding floods which resulted therefrom, and (3) the relation of maximum flood discharges to the more frequent ones, serves as an initial indicator of the flood potential. The fact that these severe floods may occur anywhere in the state, particularly in intensively developed urban areas, offers a sober challenge to flood plain managers. The purpose of this chapter is to expand upon three related topics: the flood potential; future flood problems; and the selection of design discharges for regulatory or project purposes, particularly as each relates to planning programs for flood plain management. The selection of a specific design discharge, for either regulatory or project design purposes, can be made only after a careful study of the flood potential of the area concerned and of the many implications which the proposed land uses may offer. The need for experience, foresight, and judgment will be illustrated throughout this discussion.

J. WM. FUNK is Chief Engineer for the Iowa Natural Resources Council, Des Moines, Iowa.

## STORM TRANSPOSITION

In the continental type of climate, characteristic of the Midwest and including Iowa, there is no meteorological reason why severe storms cannot occur anywhere within the region. This permits storm transposition concepts to be used.[4] The hydrologist can transpose recorded storm patterns to other river basins for which the flood potential is desired and evaluate the magnitude of flood discharge which might occur. A second approach, if one accepts the concepts of storm transposition, involves a study of the experienced peak discharges from storms of similar severity. As outlined previously, drainage area and other physiographic variables may be correlated to the magnitudes of peak discharge.

Either method makes it imperative that we obtain a record of the precipitation which occurs during severe storms. The official climatological data network of the U.S. Weather Bureau, now part of the Environmental Science Services Administration, is not sufficiently dense to record these scattered infrequent storms in great detail. The areal density of precipitation gages in Iowa is approximately one per 250 to 300 square miles. As a result, supplementary rainfall surveys ("bucket" surveys) must be made for each severe storm experienced in the state.[4] These usually are conducted whenever a maximum unofficial amount of at least 7 to 9 inches in 24 hours or less is reported by the news media or other sources. These bucket surveys are coordinated among the Iowa Natural Resources Council (INRC) and several state and federal agencies, including the Weather Bureau, Geological Survey (USGS), Corps of Engineers, and Soil Conservation Service. Because stream discharge determinations as well as flood damage observations are also desired, a great deal of activity by field and office personnel takes place during and following these severe storm periods.

## A TYPICAL ANALYSIS

The severe storm and flash flooding in the Wapsinonoc Creek basin on June 6–7, 1967, can be used to illustrate storm transposition concepts and data collection problems. The water surface profile and peak discharge data for this flood were reported in a recent USGS bulletin.[11] Since there are no official Weather Bureau precipitation gages in the basin, a bucket survey was conducted to obtain data on amounts and durations of heavy rainfall. One resident reported that a previously empty, 5-gallon bucket overflowed from direct rainfall!

**FLOOD POTENTIAL AND FUTURE FLOOD PROBLEMS** 39

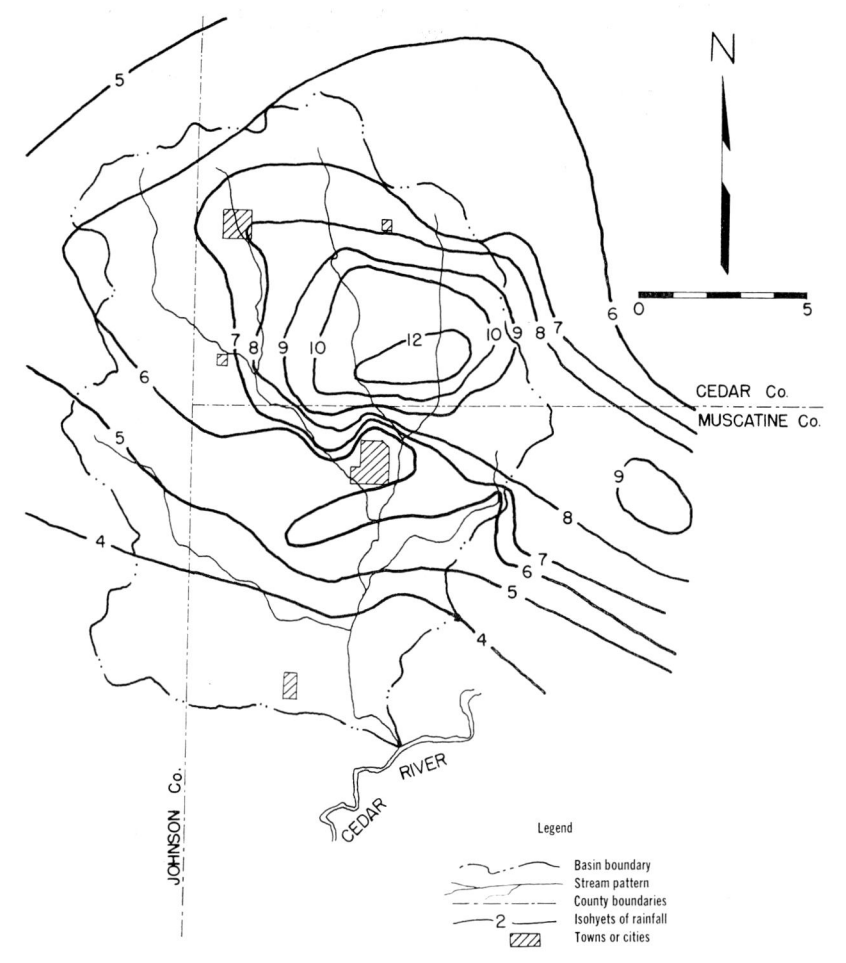

Fig. 4.1. Isohyetal map for June 6–7, 1967, Wapsinonoc Creek, Iowa. (See also storm center no. 6, Fig. 3.2.)

The isohyetal map for this storm is shown in Figure 4.1. The rain fell in a period of about 14 hours, but was concentrated in two or three rapid bursts of two to three hours each during this period. The computed depth-area rainfall data, as determined from the isohyetal map, are listed in Table 4.1.

### Other Factors

This storm produced flood peak discharges that were outstanding in eastern Iowa hydrologic experience. The sheet and gully erosion

Table 4.1. Depth-area rainfall relationship for the storm of June 6–7, 1967, Wapsinonoc Creek basin, Iowa

| Rainfall Amount, Inches | Area Over Which the Designated Amount Occurred, Square Miles |
|---|---|
| 13 | (Maximum point value) |
| 12 | 4 |
| 10 | 16 |
| 9 | 30 |
| 8 | 65 |
| 7 | 98 |
| 6 | 190 |

Source: Compiled by Iowa Natural Resources Council, Des Moines, Iowa.

and ensuing deposition were most severe, as cultivated fields had little or no vegetative cover in early June. Iowa City, which has experienced flash-flood problems on several small streams, was at the edge of the 5-inch isohyet. If this storm pattern were transposed to the Iowa City area and centered over the flash-flood problem areas, the transposition would in all probability yield a flood of catastrophic dimensions. If this storm were to occur downstream of a major flood control dam and reservoir, such as Coralville Reservoir or Red Rock Reservoir, the runoff from local uncontrolled areas could easily cause a flood of sizable magnitude, even on a major stream. For instance, newspaper accounts of the 1905 Bonaparte storm (described in the previous chapter) reported that the local flood runoff in the Keosauqua-Bonaparte area brought the Des Moines River to a standstill, and at Keosauqua the crest flood stage was within 4 feet of the record stage of 1903.

These very unusual floods also give a misleading trend to the array of flood data for a stream which includes only one of them. The tabulation of annual flood peak discharges for Davids Creek near Exira, listed in Table 4.2, illustrates this effect, which complicates the flood frequency analysis. However, in flood plain management work a problem of equally serious consequence is the array of flood data which includes some sizable floods but no severe or unusual flood events. If during the period of record these severe storms have always occurred elsewhere in the region, then no unusual flood will have been experienced. It is much more difficult under these circumstances to convince local planning agencies that the total flood experience of a region should be considered in establishing flood plain regulations.

**FLOOD POTENTIAL AND FUTURE FLOOD PROBLEMS**  41

Table 4.2. Annual flood peak discharges for Davids Creek, Iowa, indicating the comparative magnitude of the 1958 flood

| Water Year | Peak Discharge, cfs | Water Year | Peak Discharge, cfs |
|---|---|---|---|
| 1952 | 860 | 1960 | 900 |
| 1953 | 558 | 1961 | 620 |
| 1954 | 199 | 1962 | 1,550 |
| 1955 | 378 | 1963 | 322 |
| 1956 | 574 | 1964 | 2,010 |
| 1957 | 1,160 | 1965 | 1,800 |
| 1958 | 22,700 | 1966 | 3,020 |
| 1959 | 240 | 1967 | 1,380 |

Source: Data obtained from USGS publications, including 1966 revisions.
Note: Drainage area, 26.0 sq mi; average discharge, 10.5 cfs.

## VARIATIONS IN PRECIPITATION

Next let us consider the flood potential from a precipitation viewpoint. About 25 inches of rainfall is the probable maximum precipitation that might occur over a 10-square-mile area during a six-hour period in Iowa.[21] This is about twice the depth of rainfall produced by the storm which caused the very outstanding flood event on Wapsinonoc Creek, and the six-hour duration is a shorter period. In other words, much greater amounts of rainfall may occur in the future than were experienced in the Wapsinonoc Creek basin.

However, to be more realistic in hydrologic design and in economic appraisals, we normally consider storms which have a greater probability of happening. In central Iowa the precipitation amount for a 100-year recurrence interval and a duration of six hours is about 5 inches, and for a 24-hour storm about 6.5 inches for the same frequency.[20,21] A rainfall of a 100-year recurrence interval is admittedly an infrequent event, having a 1 percent chance of occurring in any single year. But the severe storms of record produce rainfall amounts which greatly exceed the magnitudes estimated for a 1 percent annual chance of occurrence. It is evident that even for a design storm of a 100-year recurrence interval, which is normal practice when economics is considered in project planning, the design storm will be exceeded sometime, someplace. It might be at the project site.

In addition, man has at the present time no control of the meteorological factors causing these severe storms. The runoff, however, will occur in a very definite pattern; its spatial and temporal variations can be estimated, and it is partially controllable by man through engineering works. These severe storms and the estimated runoff therefrom permit water resource projects to be reviewed from the standpoint of project efficiency, and the residual flood hazard and associated damages can be determined.[14,15,16] Decisions can then be made concerning the magnitude of peak discharge to be used in design or in regulation.

## CATEGORIZING PEAK DISCHARGE MAGNITUDES

When a storm occurs over a specific area, the ensuing volumes and rates of runoff will depend upon the characteristics of both the storm and the drainage basin. For some runoff events, snowmelt will also be involved. Therefore, the magnitudes of peak discharge will vary greatly from lesser floods to the great floods. Descriptive definitions of floods are needed for planning and design. The following are basic definitions used by the Corps of Engineers and accepted as levels of design criteria in categorizing the flood potential and planning and designing flood control improvements:[4,9,14,15,16]

1. The *probable maximum flood* is an estimated or hypothetical flood that represents the peak discharge that may be expected from the most severe combination of critical meteorological and hydrological conditions that are considered reasonably probable of occurrence in the region. The probable maximum flood is invariably much greater than the maximum flood of record. It is used specifically as a criterion in spillway design and to ensure the safety of dams, and represents the probable physical upper limit of the flood potential.
2. The *standard project flood* is an estimated or hypothetical flood that may be expected from the most severe combination of meteorological and hydrological conditions that are considered reasonably characteristic of the geographical region involved, excluding the extraordinary rare combinations. It has a reasonable probability of occurring and is comparable to some of the great floods which have been reported herein. It may approach or exceed approximately one-half the estimated peak discharge for the probable maximum flood condition.
3. The term *design flood* refers to the flood hydrograph or peak discharge value adopted as the basis for regulatory activities of flood plain management and the design of flood protection works or other flood plain projects. It is selected following full consideration of the flood potential; flood characteristics; and flood frequencies, economics, and other practical considerations.

# FLOOD POTENTIAL AND FUTURE FLOOD PROBLEMS

4. The term *intermediate regional flood* has been used in one recent study to define the recommended design or regulatory level.[15]

## Relationship of the Defined Floods to Experienced Floods

Floods in the first two categories have peak discharges greater than those to which a frequency or probability of occurrence can be assigned or determined through statistical analysis. The design flood, as selected for a specific flood plain management project, may be a discharge for which the frequency can be calculated, or it may be as great as the standard project flood. Evaluation must be made of the flood hazard risk before the design flood discharge can be selected from the range of peak discharges which have a reasonable potential of occurring in the future. The magnitudes of the flood discharge for these various categories at selected locations in Iowa are listed in Table 4.3.

## FLOOD DISCHARGE-FREQUENCY ANALYSIS

In the case of a specific flood plain management project, the next part of the hydrologic study process is to collect the historical flood

Table 4.3. Comparison of experienced flood discharges with discharges computed for other flood categories

| River | Drainage Area (sq mi) | Historical record | Estimated 50-year flood | Standard project | Spillway design |
|---|---|---|---|---|---|
| Des Moines River, Red Rock Reservoir | 12,500 | 155,000 | 113,000* | 273,100 | 613,000 |
| Des Moines River, Saylorville Reservoir | 5,840 | 60,200 | 59,000* | 115,400 | 277,800 |
| Skunk River, Ames Reservoir | 315 | 8,630 | 14,000 | 50,100 | 91,800 |
| Davids Creek, Exira Reservoir | 57 | 22,700† | 11,000 | 44,000 | 97,000 |

Source: Data obtained from reports of Corps of Engineers and USGS.
* Computed in Iowa Highway Research Board Bulletin 28.
† Peak discharge near Hamlin, 26.0-sq-mi drainage area.

data from various sources and study and analyze the data statistically in relation to the descriptive definitions just given. The primary source of information is the USGS, the basic data-collection agency for streamflow. Because of the need for economic analysis in most projects, a frequency approach is used in interpreting the flood discharge record. The USGS in its data analysis section uses the index flood method (mean annual flood), more commonly called the Dalrymple method, for the analysis of gaging-station data.[6] Flood frequency curves are developed for individual gaging stations, with appropriate adjustments for different lengths of record. Regional correlation of flood discharge data is accomplished through multiple correlation methods using linear regression.[2,3] Two relations are used to define the flood frequency curve at any site within a region. The first relation associates size of the index flood to selected physiographic characteristics of the basin. The second relation yields a ratio to the index flood for a selected recurrence interval, with which the desired peak discharge is computed. These techniques were used in a recent study of the magnitude and frequency of Iowa floods.[12]

The Corps of Engineers uses the Beard statistical method based on a log-normal distribution of the peak discharges.[1] This is a modification of the Pearson Type III frequency-analysis method.[7] In analyzing rainfall frequencies the Weather Bureau used the extreme value probability method of Gumbel.[20,21] This same technique is used indirectly by the U.S. Soil Conservation Service, since their runoff prediction techniques are based upon rainfall-runoff relationships. However, the Hazen method is used in peak discharge analysis.[7]

It is not surprising that differences occur in flood frequency curves computed for a specific site, with these different techniques of analysis being used. Although these differences may not be appreciable for the very common events, less than 5-year or 10-year frequencies, they may be quite sizable at the recurrence intervals commonly used for project design. This makes a realistic economic analysis difficult, and the differences frequently present some embarrassing and confusing situations for project sponsors. The flood management problems cited in many publications not only point out the location of the real residential damage potential but also emphasize the need for realistic analytical techniques for floodflow data.[13,17] Since the techniques of the federal agencies doing statistical analyses have produced quite different results, the INRC has urged the new federal Water Resources Council to adopt a uniform frequency-analysis technique, which could appropriately be used by most federal and state agencies. The Water

Resources Council has recently adopted and published a uniform analytical technique for determining floodflow frequencies.[19] This bulletin illustrates the use of the log-Pearson Type III distribution in flood discharge frequency analysis. The INRC has not included this technique in any of its official state policies as yet. Although flood damage–frequency relation data do emphasize the importance of frequency, it should be pointed out that in the range of flood frequencies most important in total flood damages the different techniques of analysis usually produce only minor differences in annual flood damage estimates.

## THE INRC DESIGN FLOOD FOR REGULATORY PURPOSES

However, there are other considerations than economics to be considered by a regulatory agency in flood plain management. These include the potential loss of life, catastrophic regional flood losses, project safety and efficiency, access during floods, and other concepts. As a result the INRC has modified the frequency approach to runoff and peak discharge prediction. For purposes of the regulation of construction in or on flood plains, an approach based upon experienced floods in Iowa has been taken.[10] This avoids the complications resulting from the several methods of frequency interpretation.

### Method of Analysis

The maximum flood peak discharges of record in Iowa were plotted, with discharge and drainage area being the selected variables. Since some of these floods are quite likely to be rare events, an envelope type curve was drawn for regulatory purposes including most but not all the events. However, a single-curve approach is not completely satisfactory because of differences in other drainage basin and storm characteristics. The state has now been divided into two areas based primarily on the glacial influence upon topography and physiography in Iowa. Two envelope curves (but not completely enveloping) for the experienced flood events from each area have been drawn. For a specific project, various flows including the design flow are presented on the experienced flood curve sheet without including a specific regulatory curve, as shown in Figure 4.2. This analysis was made for a study of Duck Creek at Davenport.[9,14]

However, the general approach does not work well in small drainage areas where individual watershed characteristics are much more

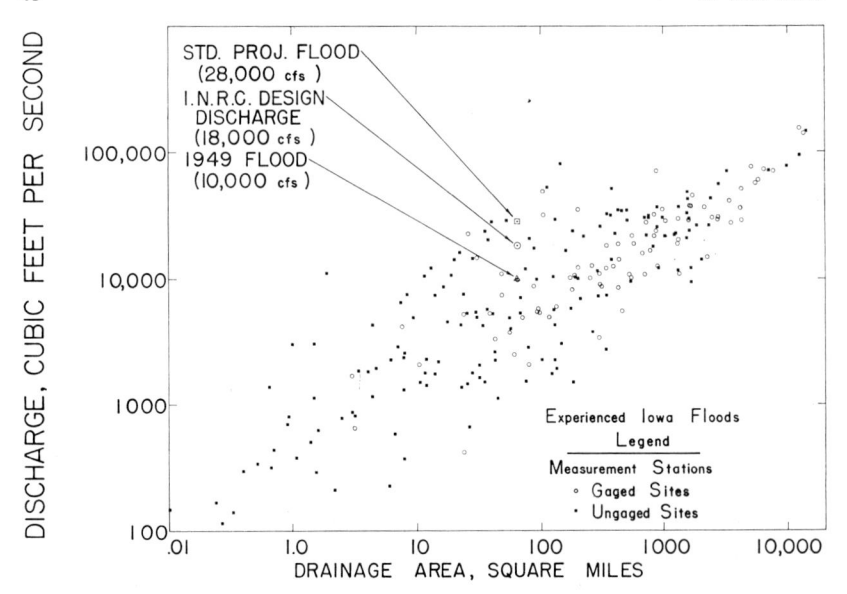

Fig. 4.2. Relation of the design flood for Duck Creek, Scott County, Iowa, with experienced floods.

important. In this situation the hydrograph prediction method of the Soil Conservation Service is quite appropriate.[18]

## Other Problems

Since the design flood for regulatory purposes is a great flood, its actual magnitude may be sensitive to the amount of valley storage. If projects exist on a large scale that severely decrease valley storage, the question is then raised whether the design discharge computed under natural conditions might need to be increased in value. This is a problem of considerable magnitude in river basin planning of comprehensive water resources programs.

When concerned with an extensive levee project which would significantly reduce valley storage, the present policy of the INRC requires a reasonable setback and alignment and a project backwater amount of 1 foot or less for the design discharge. Also, individuals in Iowa have built many rural levees which are not part of a comprehensive system, or belong only to a very limited system. Proper approval has not been obtained for many of them. If they are reviewed by the INRC, the present policy is to assume an equal and opposite

# FLOOD POTENTIAL AND FUTURE FLOOD PROBLEMS 47

encroachment and to limit the height either to the water surface elevation corresponding to the USGS 15-year frequency flood discharge, with the project in place, or to a project backwater effect of 1 foot, whichever is lower in elevation. For those levees built quite close to the streambank, the 1 foot of backwater usually controls. This means that natural valley storage will be available on most rural flood plains for floods larger than the 15-year events.

To date, the present flood plain management program has not encompassed long reaches with appreciable reduction of valley storage. As a result it has not been necessary to evaluate the effect of storage reduction on the design discharge used to establish minimum protection levels. This may not be true in the future, so more advanced flood routing methods need to be introduced.[4]

However, additional study is needed for comprehensive planning of entire river basins. In a field of engineering practice having such dynamic response and effects, one conclusion easily reached is the desirability of project sponsors to counsel with the appropriate regulatory agency very early in the design procedure for any project—flood plain management, regulation, protection works, or miscellaneous construction activities.

## SUMMARY

As we involve ourselves with future flood problems, what is the areal extent of our concern? It has been estimated that in the United States the flood plains of the rivers and streams amount to 6 percent of the total land area.[17] Since most urban developments of any size involve at least one stream or river, these developments will also involve the flood plain. If we realize that flood plain management and flood damage prevention is primarily an urban problem, the question arises as to how much of the flood plain is involved in urban development. It has been estimated that 5 percent of the Des Moines River flood plain is occupied by urban areas.[8,13] Although this is a small proportion, which in turn is only a very small part of the drainage area of the basin, nothing has been said concerning the number of people involved or the amount of investment relative to total investment in real and personal property. When the comprehensive basin framework plans for both the Upper Mississippi River and Missouri River are completed in the near future, better estimates of area both of total and of urban flood plains should be available for

Table 4.4. Residential damage data for the nation's flood plains

| Flood Recurrence Interval, Years | Homes Involved, Percent of Total | Residential Damage, as Percent of Total |
|---|---|---|
| 1–10 | 2 | 65 |
| 10–50 | 4 | 30 |
| greater than 50 | 2 | 5 |
| not flooded | 92 | 0 |
| Total | 100 | 100 |

Source: Data from a study by Marion Clawson, 1966.

the state on a major intrastate stream basis. However, these data still will not be sufficiently detailed for specific project design purposes.

In view of the low percentage of flood plain lands in comparison to total land area, it is logical to inquire how much urban damage actually occurs from flooding. Flood damage data for the United States were reported at a recent interstate conference on water problems.[5] A summary is included in Table 4.4.

The percentage of homes in flood plain areas subject to flood damages is not great, being 8 percent of the total. However, this 8 percent is much greater than the percent of total land area occupied in urban flood plains, reported as .3 percent.[17] The same trend would in all probability apply also to commercial and industrial property. This clearly indicates the greater than average pressure for urban development of the flood plains. This growth and occupancy trend, when compared with a realistic appraisal of the flood potential and associated flood risk, confirms the need for positive action through flood plain management. The concepts and means of evaluating the flood potential which were discussed herein should aid in providing flood plain managers with the technical expertise they will need in solving these development problems.

**REFERENCES**

1. Beard, L. R. *Statistical methods in hydrology.* Sacramento: U.S. Army Eng. Dist. 1962.
2. Benson, M. A. *Evolution of methods for evaluating the occurrence of floods.* USGS, WSP 1580-A. Washington: USGPO. 1962.
3. ———. *Factors influencing the occurrence of floods in a humid region of diverse terrain.* USGS WSP 1580-B. Washington: USGPO. 1962.
4. Chow, Ven T., et al. *Handbook of applied hydrology.* New York: Mc-Graw-Hill. 1964.
5. Clawson, M. *Insurance and other programs for financial assistance to flood victims.* Summary of Proc., Interstate Conf. on Water Problems. Scottsdale, Arizona. Dec. 1966.

# FLOOD POTENTIAL AND FUTURE FLOOD PROBLEMS

6. Dalrymple, T. *Flood frequency analyses.* USGS WSP 1543-A. Washington: USGPO. 1960.
7. Inter-Agency Committee on Water Resources, Subcommittee on Hydrology. *Methods of flow frequency analysis.* Bull. 13. Washington: USGPO. 1966.
8. Iowa Natural Resources Council. *An inventory of water resources and water problems, Des Moines River basin, Iowa.* Bull. 1. Des Moines. 1953.
9. ———. *Engineering report on Davenport's proposed Duck Creek development plan.* Mimeo. Des Moines. 1967.
10. Rich, E. *Study of regional floods in Iowa.* Unpublished paper presented at the County Engineer's Hydraulics Short Course, Iowa State Univ. Mimeo. INRC, Des Moines. 1966.
11. Schwob, H. H. *Flood of June 7, 1967, in the Wapsinonoc Creek basin, Iowa.* USGS open-file rept. Mimeo. Iowa City. 1968.
12. ———. *Magnitude and frequency of Iowa floods.* Iowa Highway Res. Board. Bull. 28, pts. 1 and 2. Ames. 1966.
13. U.S. Army Corps of Engineers. *Flood control, Upper Mississippi River comprehensive basin study.* App. I. Mimeo. Chicago: North Cent. Div. 1967.
14. ———. *Flood plain information report, Duck Creek, Scott County, Iowa.* Rock Island: U.S. Army Eng. Dist. 1965.
15. ———. *Flood plain information report, Cedar River, Linn County, Iowa.* Rock Island: U.S. Army Eng. Dist. 1967.
16. ———. *Interim report on flood control for Exira, Iowa.* Omaha: U.S. Army Eng. Dist. 1965.
17. U.S. House of Representatives, Committee on Public Works. *A unified national program for managing flood losses.* Rept. of the Task Force on Federal Flood Control Policy. House Document 465, 89th Cong., 2nd sess. 1966.
18. U.S. Soil Conservation Service. *Hydrology handbook.* Mimeo. Washington. 1957.
19. U.S. Water Resources Council. *A uniform technique for determining flood flow frequencies.* Bull. 15. Mimeo. Washington. 1967.
20. U.S. Weather Bureau. *Rainfall intensity-duration-frequency curves.* Technical Paper 25. Washington: USGPO. 1956.
21. ———. *Rainfall frequency atlas of the United States.* Technical Paper 40. Washington: USGPO. 1961.

Coralville Reservoir, Iowa River. Courtesy Corps of Engineers, Rock Island District.

**Part 3**

*. . . neither flood protection nor land-use regulation is a complete or adequate measure when applied alone.*

Gilbert F. White

# ELEMENTS OF A FLOOD PLAIN MANAGEMENT PROGRAM

# 5

# TECHNIQUES FOR DEVELOPING A COMPREHENSIVE PROGRAM FOR FLOOD PLAIN MANAGEMENT

Merwin D. Dougal

THROUGHOUT THE WORLD, floods periodically have caused tremendous economic losses and untold human misery and suffering, including loss of life. It is immaterial whether one first mentions Florence, Italy, and the River Arno, the lower Rio Grande in Texas, or the Mississippi River along Iowa's eastern border, for each brings to memory the sad experience of recent and severe floods. A recent bulletin of the Environmental Science Services Administration (ESSA) states:

> The transformation of a tranquil river into a destructive flood occurs hundreds of times each year, in every part of the United States. Every year, some 75,000 Americans are driven from their homes by floods; on the average, 80 persons are killed each year. These destructive overflows have caused property damage in some years estimated at more than $1,000,000,000. Floods are also great wasters of water, and water is a priceless national resource.[26]

Yet floods are natural events, a part of the life of a river system. In Iowa the historic floods of maximum severity provide an excellent

MERWIN D. DOUGAL is Assistant Professor of Civil Engineering, Iowa State University, and Conference Supervisor of the annual Water Resources Design Conference.

53

measure of the response of our river systems to severe hydrologic conditions. The elaboration which was made of the flood potential in relation to flood plain occupancy serves as an additional warning to flood plain managers. The envelope curves, based upon the historic flood record, provide a clear picture of the regional experience in Iowa. Nonetheless, the probability of yet more severe floods cannot be discounted. The destructive flood of the River Arno which ravaged Florence, Italy, in November, 1966, gained the unique distinction of being the greatest in 900 years. The maximum flood stage was more than 2 feet higher than the historic record inundation of 1333. In flood hydrology, the unusual all too frequently becomes the experience. The enormity of the problem should therefore be quite clear to those who manage the flood plains, with the added knowledge that quantitative estimates of the flood potential can be provided through hydrologic design. At this point it is appropriate to pose several pertinent questions which will serve to guide the discussion of programs for flood plain management which follows.

What will happen when future floods occur? How can we cope with the problem? What are the opportunities for flood plain management? What are the essential elements of a comprehensive program?

## PHYSICAL ALTERNATIVES

It is obvious that the works of man constructed in the flood plains are periodically subjected to flood damage. These include homes, businesses, industries, agricultural enterprises, and transportation routes and facilities. All too frequently, however, we fail to recognize that these same works can also affect the magnitude of flooding. Poor land management, as well as increased urbanization, can increase the volume and rapidity of runoff delivered to the river system. Construction of various works in the flood plain may unduly restrict the discharge, causing an increase in flood stages and attendant damages. Flood profile studies which have been completed at Cedar Rapids and Waterloo, Iowa, clearly indicate the effect caused by constrictions.[18,22] Flood stages during severe floods would be 2–4 feet higher in the upstream reaches of urban areas compared to those experienced in the more open rural reaches of the river. This effect is illustrated in Figure 5.1. Because of the interrelationship of man's works and floods, one upon the other, careful and detailed planning of the entire river basin is necessary if optimum use of the flood plain is to be achieved.[3,13]

# DEVELOPING A COMPREHENSIVE PROGRAM

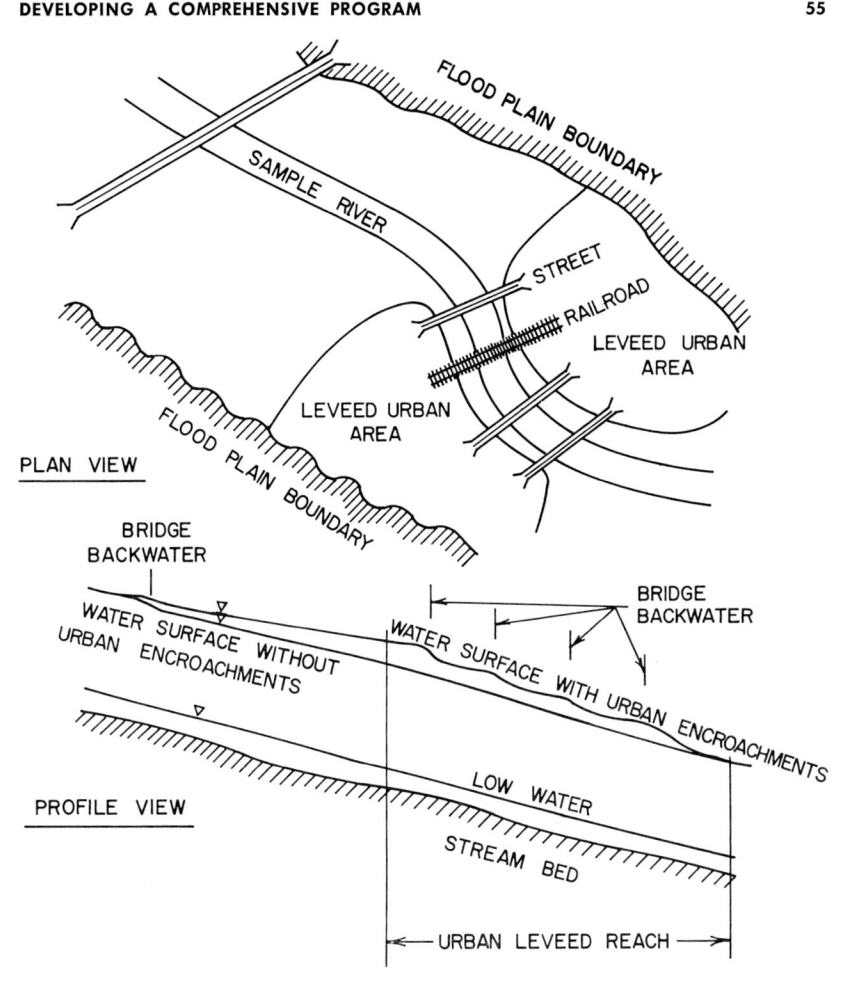

Fig. 5.1. The general effect of urban encroachment and flood plain occupancy upon flood stages.

But what can be done at the local level to solve flood problems? Either we can attempt to "control the flood waters" or we can attempt to "control use of the flood plain" or we can use a combination of the two techniques. Flood plain management encompasses both, combining proper land-use planning and regulation with flood control measures.

Although these techniques are simple in principle, complexities arise when attempts are made to apply them. This occurs because application of either technique affects the individual or corporate rights of various persons and groups in our complex industrial society; conflicts of interest abound. Technical alternatives must be weighed in relation to the socioeconomic and institutional factors which also exist.[15,16] Optimum utilization of the flood plain may be limited by one or more of these factors which becomes a constraint on the others. However, only the technical alternatives related to development of flood plain management programs will be considered herein. Formulation of this development sequence will provide a framework from which positive community action can progress.

## DEVELOPING A FLOOD PLAIN MANAGEMENT PROGRAM

What can a local community do if it faces a persistent flood problem? What can it do if a potential threat is confirmed by engineering studies, but there is no record of floods? The solution in either case is found in the following sequence of steps for formulating an action program (See Fig. 5.2.):

1. Recognition of the flood hazard by the community, its officials, and its citizens.
2. Implement and maintain an appropriate flood forecasting and warning system.
3. Develop a detailed operating schedule for flood fighting and emergency measures.
4. Outline a program for immediate adjustments in structures and occupancy in flood hazard areas.
5. Implement flood plain regulations, using (a) immediate, short-range, stop-gap measures and (b) long-range measures based upon comprehensive planning.
6. Plan for the optimum utilization of the flood plain, and implement the plans with (a) technical studies, (b) socioeconomic studies, and (c) legal-institutional-management studies.
7. Construct engineering works for the control of flood waters that are a part of the comprehensive plan for flood plain utilization and are economically feasible.
8. Operate and maintain the entire flood plain management program.

Within this framework for an action program there are both (1) direct and indirect methods of reducing flood damages by combating or controlling floodwaters and (2) direct and indirect methods of controlling the use of land and structures in the flood plain. Use of these methods will be discussed in additional detail, with major em-

## DEVELOPING A COMPREHENSIVE PROGRAM

Floods occur or flood hazard determined through studies
↓
Gain recognition of the flood hazard
↓
Implement flood forecasting and warning system
↓
Develop operating schedule for flood fighting measures
↓
Make immediate adjustments in structures and occupancy
↓
Formulate, implement flood plain regulations
↓
Plan for optimum future use of the flood plain
↓
Complete necessary construction programs which accompany long-range plans
↓
Operate, maintain, up-date the action program for flood plain management

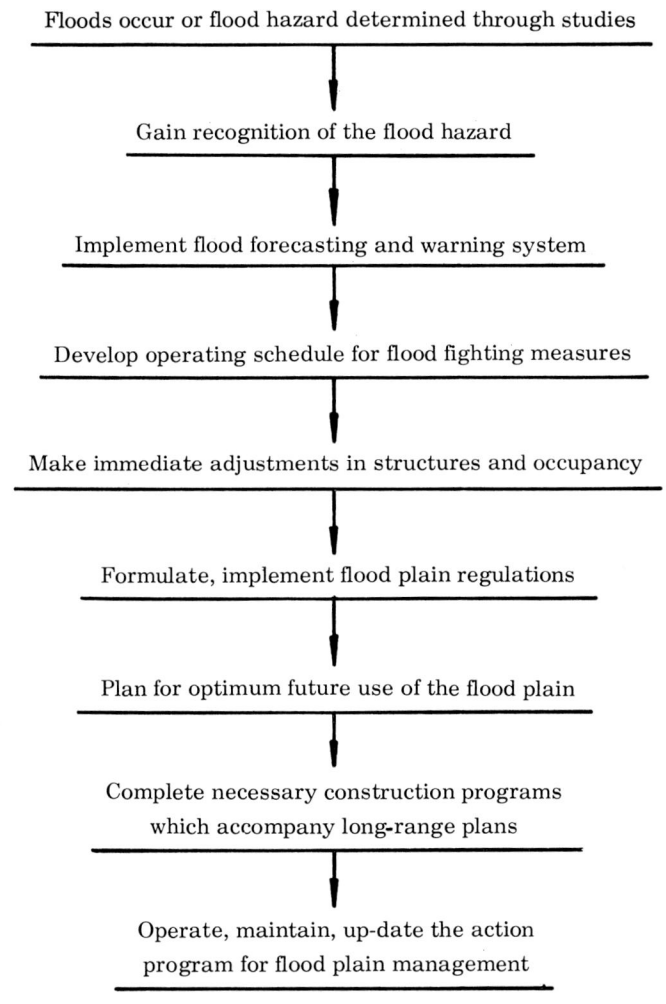

Fig. 5.2. Steps in the formulation of a flood plain management program.

phasis upon the former to permit the remainder of the book to be devoted to land-use planning and regulation concepts.

## RECOGNITION OF THE FLOOD HAZARD

All too frequently, the knowledge of specific flood hazard areas is gained through experience; people's minds become dulled as time passes, and the next flood catches the area as unprepared as did the first one. Two prevalent attitudes toward the flood hazard were reported in recent studies.[17,31] One is the belief that although there may be some vague recollection of past damage, a flood hazard no longer exists. A second attitude, common among the owners of new developments on the flood plain, is that the advantages are worth the risk. Some of these subdividers and developers, it was noted, were aware that they faced the flood risk for a brief period only; thereafter, it belonged to the new owners. Hazard to loss of life by either flash flooding or structural failures of partial protection works are seldom considered in such evaluations, yet tragic losses of life have occurred when flooding became a reality.[7,14,24] Owners of new developments are not necessarily elated with the knowledge gained through flooding; and in one study, owners of new commercial buildings reported they would not have invested in flood plain locations if the hazard had been known to them.[31] However, some residents have an affinity for the flood plain, as indicated in a study by Roder which showed that a large percentage of flood plain occupants previously had been flooded elsewhere.[17] The incentive to remain in a flood plain area was strong with certain individuals and groups who chose, however, to ignore or evade questions regarding their knowledge of the existence of a real flood hazard. In view of these many facets, the problem of initiating and maintaining community awareness of a flood hazard will require continued and concerted action by alert individuals, dedicated civic groups, and responsible public officials.

In areas where frequent and persistent flooding occurs, there should be less difficulty in obtaining and maintaining recognition of the hazard. However, in such a case, the problem may be so perennial that general apathy exists toward developing an action program of any type. This is considered to be true especially in areas where the so-called underprivileged segment of society is located in flood plain areas due to a combination of events, including the effect of floods in creating substandard housing conditions. Residential occupants of the flood plains of large rivers in major urban areas are notably in the lower income and education groups, with less skilled occupation specialties.[17,31] This is not true, however, of newer residential developments in tributary valley areas, nor of commercial and industrial enterprises in either type of area.[2,17]

An even more difficult task confronts the community which becomes aware of the flood potential through engineering studies but is without the benefit of a major flood experience. In the minds of the unwary, placid streams are not supposed to become violent and angry and to overflow their banks. The availability of engineering reports

## DEVELOPING A COMPREHENSIVE PROGRAM

outlining the hazard does not in itself suffice to guide the flood plain manager. At Carnegie, Pa., some property owners were not aware of the study which had been made.[31,32] At Topeka, Kans., local residents and businessmen could not interpret the information contained on flood maps, or they erroneously believed that the hazard had been alleviated by protection works since the mapping program had been completed.[17] In Iowa, flood plain information studies have been made for at least two tributary streams in urban areas where severe flooding has not been experienced, but where intensive development for urban uses is rapidly progressing.[8,22,23] In these instances a real challenge is presented to community leaders in achieving and maintaining a positive awareness of the flood hazard.

Between these two extremes, frequent flooding versus none, one can find the remainder of people's knowledge of the hazard. Although flood hazard evaluation techniques are well developed, techniques for obtaining widespread use and understanding of published results are lacking.[1,3,11] Regardless of the location or degree of the flood hazard, common recognition is the first step and the key to an action program for flood plain management in any community. The remaining steps in formulating such a program can then be accomplished with greater ease and with increased opportunity for success.

### FLOOD FORECASTING AND WARNING SYSTEMS

The federal government since 1891 has assumed major responsibility for flood forecasting in the United States.[28] The immediate purpose of the program has been the reduction of damage and loss of life. The national approach is desirable in view of the interstate character of most river basins and the advantages of a coordinated network for effective operation. The primary responsibility for this service is assigned to the U.S. Weather Bureau of the ESSA. Forecasts of initial flood conditions and crest stages are provided to local communities in addition to forecasts of other meteorological information.

Regional river forecast centers have been established by the Weather Bureau and currently serve over three-fourths the conterminous United States.[26,30] In the Midwest, the Missouri River basin above Kansas City is served by the Kansas City River Forecast Center. The Upper Mississippi River basin is served by the St. Louis River Forecast Center. The river basin area served by a center is divided into several districts. One Weather Bureau station is selected in each as a river district office. The Des Moines Weather Bureau airport station is the

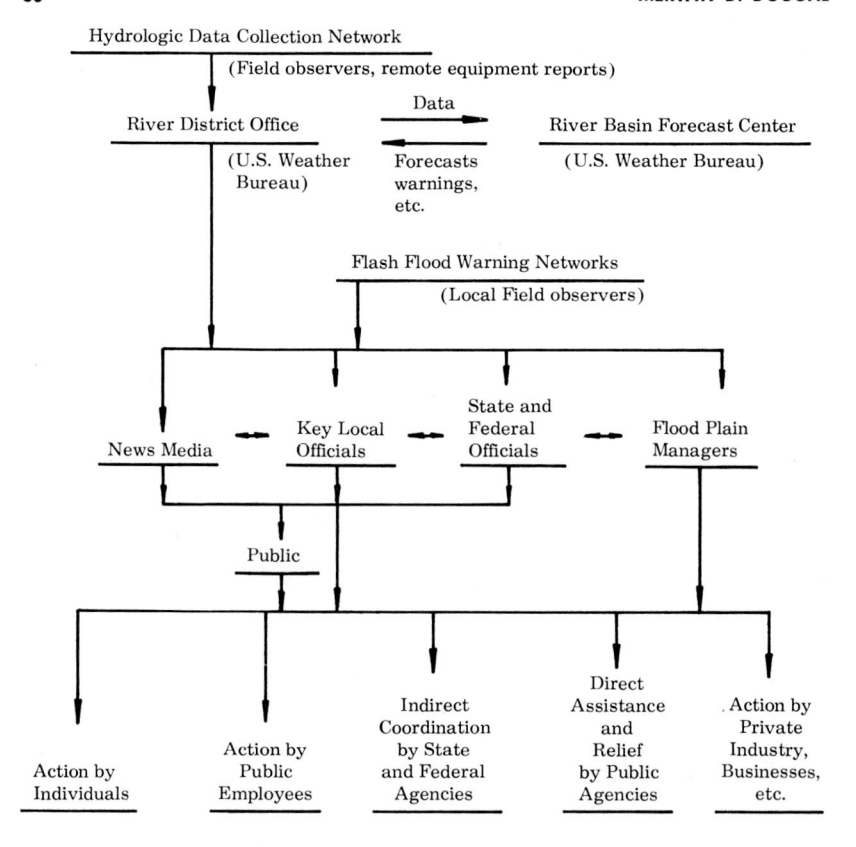

Fig. 5.3. Sequential actions in establishing an effective flood forecasting and warning system.

river district office for most of Iowa. Other district offices for smaller areas in Iowa are at Sioux City, Minneapolis, Moline, Burlington, Kansas City, and Omaha. However, operations at the smaller offices gradually are being assimilated by the large regional forecast centers.

The general pattern of flood forecasting involves several sequential steps, illustrated in Figure 5.3. The district office collects precipitation data from the climatological network and river stages at selected stream-gaging stations. These reports are relayed to the river forecast center. Hydrologists use snowmelt and rainfall-runoff relationships in conjunction with flood routing methods and from the observed data compute the predicted flood stages at various points on a stream. Anticipated weather conditions are also included. Electronic

## DEVELOPING A COMPREHENSIVE PROGRAM

computers speed the process of analysis and prediction, and weather radar instruments provide additional regional data on storm locations and movements. In Iowa several telemetering precipitation gages have been installed within 50–100 miles of Des Moines to provide more direct and rapid information on rainfall amounts in the Des Moines River district. The flood forecast is transmitted to the local river district for dissemination to the responsible local authorities and to the public by the news media. Through these sources the residents of the flood plain receive warning of impending flood conditions, and precautions or protective measures can be taken.[26,28,29]

The time lag between rainfall or snowmelt and the resulting rise of streams and rivers depends primarily upon the size or drainage area of the stream. Three classifications of streams have been made by the Weather Bureau: headwaters, tributaries, and main stem. Headwater and tributary areas which affect urban areas are of great concern because the time lag between observed rainfall and crest flood stage is very short, frequently in terms of a few hours. Swift action is necessary to save lives, and there is seldom time to protect or remove personal belongings. In Iowa a system of flash-flood warning networks has been developed and operated for over a decade in certain problem areas. The Weather Bureau and the Iowa Natural Resources Council were instrumental in establishing this system. These networks as of 1964 were operational in the following areas: Four-Mile Creek and Walnut Creek at Des Moines, Lizard Creek at Ft. Dodge, Squaw Creek at Ames, East Nishnabotna River at Atlantic, Black Hawk Creek at Waterloo, Winnebago River at Mason City, Upper Cedar River at Charles City, and the Iowa River above Marshalltown.[9] Several tributaries in the Cedar Rapids area have since been included. The Iowa network is shown in Figure 5.4.

The flash-flood warning system makes use of immediate precipitation reports from local observers, either direct to the Weather Bureau or to a local police or sheriff's office for radio communication relay. In two basins, radios are available at observer locations in case telephones are inoperative. The weather surveillance radar at the Des Moines Weather Bureau airport station also covers these areas, and strong echos observed with radar indicate areas of potentially heavy precipitation. These various rainfall reports and radar data are used to make rapid estimates of the flood potential, and flash-flood warnings are issued directly from the district office. Immediate radio and television broadcast of information is thereby possible for areas where intense precipitation and probable severe flash flooding are expected.

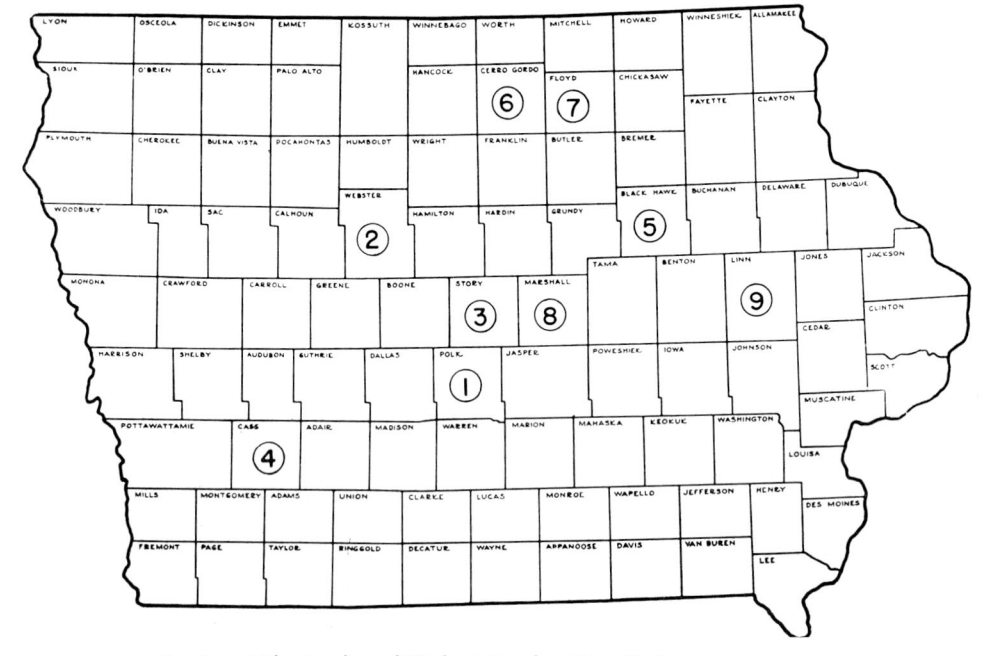

1. Four-Mile Creek and Walnut Creek at Des Moines
2. Lizard Creek at Fort Dodge
3. Squaw Creek at Ames
4. East Nishnabotna River at Atlantic
5. Black Hawk Creek at Waterloo
6. Winnebago River at Mason City
7. Upper Cedar River basin above Charles City
8. Iowa River basin above Marshalltown
9. Indian and Prairie Creeks at Cedar Rapids

Fig. 5.4. Headwater and tributary basins included in the Iowa flash-flood warning network.

Local police and other authorities are alerted to provide additional warnings and also to check on stream or river stages at key locations. In some communities a local representative is authorized by the Weather Bureau to make rapid analysis of local data, prepare a flood forecast, and issue an official flash-flood warning to the public. Because it is probable that major flash flooding will occur in late evening or during the night, warnings need to be issued to local authorities as well as to the radio and television media to arouse sleeping residents.[12]

Flash-flood warning networks as well as flood forecasting systems require 24-hour attention and continuity of operating personnel. Therefore, periodic inspections are needed to ensure that an effective

## DEVELOPING A COMPREHENSIVE PROGRAM 63

organization remains in force. Otherwise, the forecasting or warning service may fail at a crucial time. The actual cost of operating a flash-flood warning system is small if existing facilities and personnel are incorporated into the plan.[28]

Several other agencies involved in the overall aspects of flood forecasting should be mentioned. The U.S. Geological Survey is responsible for river discharge measurements and the stream-gaging program. Flood discharge measurements, rates of stage increase, and other observations of river behavior are field data needed at the forecast center to supplement historic data and observed precipitation and weather phenomena.[19] The Corps of Engineers frequently obtains flood discharge measurements in areas where federal water resources and flood protection projects are proposed or are completed and operational. The Corps of Engineers is responsible also for operation of several flood control reservoirs in or near Iowa, including the evaluation of downstream effects of reservoir flood releases. Various local, county, and state agencies are concerned if flooding is widespread and are alerted by the Weather Bureau under prearranged procedures.

Communities should make adequate provision that local residents understand the meaning of flood forecasts. Flood stage forecasts predict the water surface level at specified river gage locations. Crest stage forecasts predict both the maximum stage, or water surface level in feet that a river is expected to reach at that gage, and the approximate time of the crest stage. The fact that an arbitrary level datum is selected at each specific gage location frequently compounds the problems of understanding flood stage predictions. The zero gage datum of a river gage is selected at an elevation below streambed to avoid negative readings during low-flow periods.

The zero gage datum may not be carefully selected in comparison to upstream or downstream gages, and an unfortunate situation may develop. For an equivalent flood discharge in cubic feet per second, the flood stage at one gage location may be 20 feet, and at another it may be only 15 feet. If the latter station is downstream from the first, then an early forecast of a 20-foot crest stage at the upstream gage, if unaccompanied by a 15-foot forecast for the latter, can unduly alarm the downstream residents for whom a stage of 20 feet may represent severe flooding. This situation actually exists in the Des Moines River basin between Tracy, Eddyville, and Ottumwa. An optimum plan for a zero gage datum in long reaches of a large river should provide for the same stage reading at which general overbank flooding would occur in rural or unprotected urban areas. Forecasts of crest flood stages should include also the stages at which damages will be of concern and should consider both protected and unprotected areas. For instance rural flooding may begin when bankfull stages are reached, whereas urban areas protected by levees will have both an overtopping stage as well as a lower stage at which flood operations need to become operative.

## FLOOD FIGHTING AND EMERGENCY MEASURES

The flood forecasting and warning system can alert local residents and responsible authorities, but prompt action is necessary if damages and loss of life are to be reduced. The next step is development of a plan for emergency action. Detailed operating procedures or schedules for effective flood fighting and other emergency measures should be available for rapid implementation.

Overall public responsibility for flood fighting remains with local authorities.[21] Elective officials can delegate responsibility to various municipal or community departments and agencies. The public works or engineering department, streets and maintenance, waterworks, waste treatment, and other departments all can play an active role. Police and traffic departments can control movement of people and vehicles. The police and fire departments invariably are concerned with rescuing stranded citizens. School officials and students have coordinated in recent floods by providing volunteer labor for filling sandbags, evacuation of goods, and other tasks. Local Red Cross and Salvation Army personnel assist in providing relief measures and temporary housing.

Various state, regional, and federal agencies act in an active and/or advisory capacity. These include the State Highway Patrol, Civil Defense, governor's office, Adjutant General and National Guard, and miscellaneous state agencies concerned with water resources and floods. Federal agencies include the Corps of Engineers, Office of Civil and Defense Mobilization, Soil Conservation Service, Agricultural Stabilization and Conservation Service, Geological Survey, and others. The Corps of Engineers, responsible nationally for flood control and flood management services, has experienced personnel available who can act in an advisory capacity during emergency periods. They are especially concerned with proper operation and maintenance of local engineering works for flood control which have been constructed with federal funds.[21]

## DEVELOPING A COMPREHENSIVE PROGRAM

A detailed operating schedule must include:

1. What has to be done (a) for minor floods, (b) for moderate floods, and (c) for severe flood conditions?
2. Who is to do it?
3. How is it to be done?
4. How can the schedule be improved?

These provisions will vary for each community, depending upon the flood hazard. Knowledge of physical facilities, flood hazard areas, available personnel, equipment, and supplies is required. Typical of flood fighting operations are: patrolling river reaches, levees, and floodwalls; making gate and conduit closures; starting relief pumps for drainage; evacuation of people and/or movable goods; reinforcing or raising levees; barricading buildings; filling sandbags; controlling traffic movements; policing evacuated areas; rescuing stranded citizens; and introducing emergency health measures and inoculations.

Technically, an emergency plan for flood fighting should provide for four stages.[21]

1. *Alert stage.* Key supervisory personnel are alerted to the flood possibilities through the forecasting and warning system. Flood stages at this point are anticipated to become sufficiently high so that additional action may be required. Equipment is checked to ensure operational capability, supplies are rechecked, and operating personnel accounted for.
2. *Activation stage.* Personnel are called to duty. Equipment and supplies are loaded and dispatched to duty locations.
3. *Operation stage.* All emergency operations are carried out as scheduled. Floodgates are installed, traffic rerouted, levees and floodwalls patrolled, relief and temporary evacuation plans initiated, emergency works constructed, and other duties performed.
4. *Deactivation and reporting stage.* Following a flood, reports should be filed by all key supervisory personnel. Review of the operating schedule is made to determine if revisions are necessary. Flood cleanup measures are carried out.[25,26]

The cost of flood fighting emergency measures can be substantial. In 1965 most Iowa communities along the Mississippi River found their treasuries severely depleted, even though the amounts of flood damage prevented were tremendous. If the area is declared a disaster area under specified federal procedures, emergency funds become available for reimbursement of some of the costs of flood fighting. However, a considerable amount of paper processing with appropriate forms is required, and frequently the amount of time expended at the local and state levels is considerable. The delicate nature of coordination and advising by state and federal agencies also was brought out during the 1965 floods on the Mississippi River. Local officials pointed out that they were encouraged to initiate and implement an extensive program for emergency protective measures that required substantial expenditures. Yet no ready answer or offer of reimbursement was forthcoming from the same advisers immediately after the floods when the claims by suppliers and equipment-rental agencies poured in.

To remedy the problem, the General Assembly in 1967 enacted a disaster aid law.[5] The act provides that governmental subdivisions including cities, towns, counties, and school districts may apply, upon the occurrence of a natural disaster, to the state government for financial assistance in the form of noninterest-bearing loans. Eligibility is based upon declaration of a disaster area by the governor and showing of "an immediate financial inability to meet the continuing requirements of local government." The amount of a loan may be up to 75 percent of obligations and expenditures actually incurred in a natural disaster or expended in lessening its impact. The aggregate total of such loans is not to exceed $1 million per biennial fiscal term.

## CONTROLLING THE USE OF FLOOD PLAIN AREAS

Flood plains are simply land areas which have the uniqueness of being subject to overflow. As land areas they can be treated within the customary and accepted confines of land-use planning and regulation. In planning and regulation programs there is no reason to exclude flood plain areas either from the controls placed over other lands or from controls specifically applicable to the flood problem. A brief discussion of three steps for controlling use of flood plains will suffice at this point, since detailed attention is directed toward these aspects in the chapters which follow. First, immediate adjustments in structures and occupancy can reduce the flood damage potential. Low damage potential can be accomplished through changed building occupancy, changed use, or acquisition of the area. For example, the city of Ottumwa, Iowa, rapidly evacuated the low and frequently

## DEVELOPING A COMPREHENSIVE PROGRAM

flooded Central Addition following the severe 1944 and 1947 floods of the Des Moines River. The land, purchased by the city, has subsequently been used for transportation route extensions and for open space and park development. Sheaffer prepared an extensive report on floodproofing as a structural adjustment.[20] Both exterior wall, window, and entrance alterations and interior innovations can transform a building into a low damage-hazard category. Mobile display counters, for instance, permit goods and equipment to be moved rapidly to second- or third-floor levels. Abandoning basement storage may be a prime prerequisite in business and commercial areas for accomplishing an immediate reduction in the flood damage potential.

According to Murphy, flood plain regulations as a second control measure consist of the following: (1) statutes; (2) zoning ordinances; (3) subdivision regulations, including utility extensions; (4) building codes; (5) miscellaneous ordinances; (6) urban renewal; (7) permanent evacuation; (8) government acquisition; (9) building financing and related tax assessment adjustments; (10) warning signs and notices; and (11) flood insurance.[14] Two purposes of regulation are outlined by Murphy. The first is to maintain an adequate *floodway,* having sufficient cross-section area for conveying flood discharges by preventing flow-constricting developments in such areas. The second purpose is the regulation of development on the remainder of the *flood plain* outside the floodway limits—sometimes appropriately called the floodway fringe. These terms, as further defined in the Iowa statutes, are:

1. *Flood plain* means the area adjoining the river or stream, which has been or may be hereafter covered by floodwater.
2. *Floodway* means the channel of a river or stream and those portions of the flood plains adjoining the channel, which are reasonably required to carry and discharge the floodwater or floodflow of any river or stream.[6]

The third and most important step is planning for the optimum utilization of the flood plain. This is the most challenging task confronting flood plain managers. As in all planning endeavors, a three-dimensional framework exists between the technical, the economic, and the institutional roles.[4] Within the limitations of the various elements, we would like to obtain the maximum net economic return from use of the nation's flood plains. This would include the full consideration and inclusion of the social costs and benefits of flood plain occupancy as well as those accruing to individuals and special interest groups. Public regulation of land use would endeavor to guide flood plain uses toward this goal. The role of the various levels of government—federal, state, and local—and of individual managers of flood plain property in achieving a comprehensive flood loss management program was outlined in a recent task force report.[27]

## ENGINEERING WORKS FOR CONTROL OF FLOODWATERS

For those engineers engaged in design and construction, there is both remuneration and satisfaction in seeing extensive projects completed and placed into operation. To the public such works appear as outstanding examples of man's ability to conquer nature. However, the truth of the entire matter became evident in the 1950's when research indicated that despite an accelerated program to construct engineering works for flood control, losses continued to climb.[27,31] Man was encroaching into the flood plain at a pace so rapid that the magnitude of new flood problems exceeded that being alleviated. In the last decade flood plain management has superseded "flood control by engineering works" as the accepted method for obtaining comprehensive planning and wise use of the flood plains.

Today engineering works for flood control are a part of the flood plain management program. Economic studies determine if they are a feasible part of a specific program. In many cases, where there is no place to locate urban developments except on the flood plains, these works will be a primary part of a river basin flood plain management program. In other areas there may be no physical possibility of constructing engineering works or they may not be economically justifiable. The proper role of engineering works is within the total planning concept, including control of use of the flood plain.

The types of engineering works for control of floodwaters can be placed into five classifications:

1. Channel improvements and drainage facilities, including open channels and closed conduits.
2. Levees and floodwalls, including interior drainage and with or without channel improvements.
3. Bypasses and diversion channels.
4. Reservoirs for temporary storage of floodwaters.
5. Land management and watershed development.

Combinations of these are frequently used in comprehensive river basin planning to provide an optimum solution to the overall flood

Fig. 5.5. Channel and levee improvements on the lower reaches of the Little Sioux River (courtesy Corps of Engineers, Omaha District).

plain management problem where engineering works are required.[13]

These engineering works accomplish their purpose of reducing flood damages by one of two methods: confining the floodwaters that naturally occur or accomplishing a reduction of flood stages by use of one or more of the five types.

Channel improvements are designed to decrease the length of the river or stream and to improve its conveyance ability. The resultant increase in velocity of streamflow permits some reduction in stage to be obtained. Combined with levees and/or floodwalls, channel improvements offset the increases in flood stages caused by the

Fig. 5.6. Concrete floodwalls, levees, and gate closures at Muscatine, Mississippi River, and Mad Creek areas (courtesy Corps of Engineers, Rock Island District).

levees confining the flow. Conservationists frequently view extensive reaches of improved, straightened channels as "biological deserts," showing that reduction of flood damages through hydraulic efficiency cannot be obtained without affecting another use of the river. There are few streams in Iowa which have not been improved or straightened for at least a portion of their length.[9] Closed conduits are used frequently in urban areas for enclosing small streams.

Levees and floodwalls have been constructed extensively along all major rivers in the nation and in Iowa.[9] Floodwaters are confined to the floodway between the levees, with the protected areas behind the structures at an elevation below the flood stage. Levees, or earth embankments, are constructed where earth materials are suitable and available and where right-of-way is inexpensive. In urban areas, where

## DEVELOPING A COMPREHENSIVE PROGRAM

many industrial buildings and railroads hug the river banks, concrete floodwalls are utilized to reduce the right-of-way requirement. Tiebacks, removable gates for temporary closures, interior drainage, patrolling during floods, emergency operations for seepage, and the like, must be considered in levee and floodwall design. The inherent problem of rapid inundation of a protected area in case of failure or overtopping is a serious matter that must be included in evaluating the use of levees and floodwalls.

Bypasses and diversion channels are designed to relieve the main river channel by conveying a portion or all of the flood discharge around the protected area. An example in Iowa is the Dry Run diversion channel at Decorah. The stream originally flowed through the entire length of the municipality but was diverted through a narrow bluff directly to the Upper Iowa River as part of a federal-local flood control project.

Reservoirs for temporary storage of floodwaters are of two types. Detention reservoirs have uncontrolled outlets and are designed to fill and empty automatically, limiting the discharge in the downstream channel to a predetermined value. Storage reservoirs have multipurpose concepts in their design, and releases are controlled. The floodwater volume may actually refill storage capacity allocated to other uses and may be released days or even months later. The system of main-stem reservoirs on the Missouri River in the Dakotas is an example. By temporarily storing the floodwaters, flood stages are reduced downstream. However, because the same flood volume must be released eventually, sustained periods of bankfull flows may be experienced. Agriculture uses and river bank stabilization measures need to be considered carefully in designing and evaluating the economics of reservoirs. In Iowa, Coralville Reservoir is in operation; and Red Rock, Saylorville, and Rathburn are major reservoirs currently being constructed.

Land management and watershed development play an important role in tributary river basin planning.[10] Two types of measures are recognized today. Land treatment and soil conservation practices enhance the preservation of soil fertility and decrease erosion and sediment problems. Structural measures include erosion control and gully stabilization structures, watershed floodwater retarding structures, and related channel improvements. Flood damages are prevented by reducing flood stages and by decreasing sediment damage. Watershed planning and development has been most active in western and southern Iowa areas.

Fig. 5.7. Diversion channel and levees for a new outlet for the Floyd River at Sioux City, Iowa (courtesy Corps of Engineers, Omaha District).

These direct methods of controlling floodwaters can be utilized at the local level but more appropriately should be included in comprehensive river basin planning for all beneficial uses of water. This would include the flood plain management program. State and federal programs for planning and constructing engineering works for control of floodwaters in Iowa are the primary responsibility of the Iowa Natural Resources Council, the Corps of Engineers, and the Soil Conservation Service (for small watershed development).

## CONCLUSIONS

The steps required to implement an action program for a flood plain management program have been outlined. Basically, we either control the floodwaters or the use of the floodplains or use some optimum combination of the two. An action program provides for recognition of the flood hazard, flood forecasting, flood fighting emergency measures, short-range plans for alleviating flood damages, and long-range plans to accomplish the wise use of flood plain areas.

The indirect and direct methods of reducing flood damages through control of floodwaters have been reviewed. Flood forecasting and warning systems can be implemented swiftly at little cost to the community. Flood fighting and emergency protection measures can assist in reducing flood losses. Engineering works for flood control provide additional measures for direct and effective control of floodwaters, and permit substantial reductions of flood losses where they are economically feasible.

The implications for funding, organization, and personnel appear great if we are to accomplish all these endeavors in a modicum of time. However, existing agencies can be used at the federal level and in most states. A reorientation or strengthening of activities is a major requirement at several levels of government, but mainly at the state and local levels. Each of these measures should be considered a part of the total flood plain management program for a river basin or for acceptance by a state government.

## REFERENCES

1. American Society of Civil Engineers, Guide for the development of flood plain regulations. Progress rept., Task Force on Flood Plain Regulations. *Proc. Am. Soc. Civil Eng.*, Hydraulics Div. 88, no. HY5, Paper 3264 (Sept. 1962).
2. Burton, Ian. Invasion and escape in the Little Calumet. In G. H. White, ed., *Papers on flood problems*. Dept. of Geog. Res. Paper 70. Chicago: Univ. of Chicago Press. 1961.
3. Chow, Ven Te, et al. *Handbook of hydrology*. New York: McGraw-Hill. 1964.
4. Heady, E. O., and Timmons, J. F. Economic framework for planning efficient use of water resources. In J. F. Timmons et al., *Iowa's water resources*. Ames: Iowa State Univ. Press. 1956.
5. Iowa, Acts of the 62nd General Assembly. Ch. 93. 1967.
6. Iowa Code. Ch. 455A. 1966.
7. Iowa Natural Resources Council. *An inventory of water resources and water problems, Floyd–Big Sioux river basins, Iowa*. Bull. 4. Des Moines. 1956.

Fig. 5.8. A multipurpose dam and reservoir which includes flood control storage, Red Rock Dam and Reservoir, Des Moines River, Iowa (courtesy Corps of Engineers, Rock Island District).

8. ———. *Effects of flood plain encroachment, Indian and Dry creeks, Linn County, Iowa.* Mimeo. Des Moines. 1966.
9. ———. *Report for the biennial period, July 1, 1962, to June 30, 1964.* Des Moines. 1964.
10. Iowa State University. *A guide to a step-by-step approach in watershed development.* Coop. Ext. Serv. Pam. 265. Ames. 1960.
11. Kates, R. W., and White, G. H. Flood hazard evaluation. In White, *Papers on flood problems.*
12. Linsley, R. K., et al. *Applied hydrology.* New York: McGraw-Hill. 1949.
13. Linsley, R. K., and Franzini, J. B. *Water resources engineering.* New York: McGraw-Hill. 1964.
14. Murphy, F. C. *Regulating flood plain development.* Dept. of Geog. Res. Paper 56. Chicago: Univ. of Chicago Press. 1958.
15. Prest, A. R., and Turvey, R. Cost-benefit analysis: A survey. *Econ. J.* 75, no. 300 (Dec. 1965).
16. Renshaw, E. F. The relationship between flood losses and flood control benefits. In White, *Papers on flood problems.*

17. Roder, W. Attitudes and knowledge of the Topeka flood plain. In White, *Papers on flood problems.*
18. Schwob, H. H. *Cedar River basin floods.* Iowa Highway Res. Board Bull. 27. Ames. 1963.
19. Schwob, H. H., and Meyers, R. E. *The 1965 Mississippi River flood in Iowa.* USGS open-file rept. Iowa City. 1965.
20. Sheaffer, J. R. *Flood proofing: An element in a flood damage reduction program.* Dept. of Geog. Res. Paper 65. Chicago: Univ. of Chicago Press. 1960.
21. U.S. Army Corps of Engineers. *Flood emergency manual.* Washington: Chief of Engineers. 1959.
22. ———. *Flood plain information report, Cedar River, Linn County, Iowa.* Rock Island: U.S. Army Eng. Dist. 1967.
23. ———. *Flood plain information report, Duck Creek, Scott County, Iowa.* Rock Island: U.S. Army Eng. Dist. 1965.
24. ———. *Report on the flood of July, 1958, in the Nishnabotna River basin, Iowa.* Mimeo. Omaha: U.S. Army Eng. Dist. 1959.
25. U.S. Department of Agriculture. *First aid to flooded homes and farms.* Agr. Handbook 38. Washington: USGPO. 1964.
26. U.S. Environmental Science Services Administration. *Floods and flood warnings.* Publ. ESSA-PL 660030. Washington: USGPO. 1966.
27. U.S. House of Representatives, Committee on Public Works. *A unified national program for managing flood losses.* Rept. of the Task Force on Federal Flood Control Policy. House Document 465, 89th Cong., 2nd sess. 1966.
28. U.S. Senate, Select Committee on National Water Resources. *River forecasting and hydrometeorological analysis.* Committee Print 25, 86th Cong., 1st sess. 1959.
29. U.S. Weather Bureau. *Elements of river forecasting.* Tech. Memo. Hydro-4. Washington: ESSA. 1967.
30. ———. *The Weather Bureau and water management, role of river forecasting and hydrometeorological analysis.* Washington: USGPO. 1965.
31. White, G. H., et al. *Changes in urban occupancy in flood plains in the United States.* Dept. of Geog. Res. Paper 57. Chicago: Univ. of Chicago Press. 1958.
32. Wiitala, S. W., et al. *Hydraulic and hydrologic aspects of flood plain planning.* USGS WSP 1526. Washington: USGPO. 1961.

Squaw Creek at Iowa State University, 1954. Courtesy Ames **Daily Tribune**.

# Part 4

*The far greater number of decisions affecting new development are made by private individuals and corporations within the limits set by state and local plans and regulations.*

Task Force on Federal
Flood Control Policy

# LAND-USE PLANNING AND DEVELOPMENT METHODS FOR FLOOD PLAIN AREAS

# 6

# A BRIEF BACKGROUND
# OF THE PLANNING PROCESS

### Burl A. Parks

LAND-USE PLANNING in flood plains is a part of the overall community planning process. For this reason it is a normal function of the local planning commission. All communities with a planning commission, a zoning commission, or both have established them under the laws of the state in which they are located. The general role of the planning and zoning agencies will be described in this chapter, providing a basis for the several phases of flood plain planning discussed in the chapters which follow.

## PLANNING AND ZONING COMMISSIONS

The usual procedure is for the local legislative body, which in Iowa is the city council or county board of supervisors, to adopt an ordinance or a resolution establishing such commissions. Once this is done, members are appointed in the manner prescribed by the applicable state enabling act.

BURL A. PARKS is Assistant Professor of Landscape Architecture, Iowa State University, and Coordinator of University Extension Planning.

State laws define the duties of such commissions and describe their relationship with the local legislative body. In Iowa the basic laws which provide the authority for planning by local government bodies are as follows:

1. Chapter 373, Iowa Code (1966)—City Plan Commission.
2. Chapter 414, Iowa Code (1966)—Municipal Zoning.
3. Chapter 358A, Iowa Code (1966)—County Zoning Commission.
4. Chapter 110, Iowa Code (1966)—Regional Planning Commissions.

All planning acts are permissive in nature. The Iowa Code permits cities, towns, and counties to establish commissions and to exercise the powers granted in the act if they so desire. However, it is not mandatory that they use such laws. Several counties in Iowa, for instance, have never taken advantage of the provisions of Chapter 358A under which more intensive land uses in rural areas may be regulated. In addition to these four basic state laws, there are numerous others pertaining to water resources and flood plain development which concern local governing bodies.[6]

Land-use planning is accomplished through the provisions of Chapter 373 of the Iowa Code. The planning commission established under the provisions of this act is charged with the responsibility of preparing a comprehensive long-range plan for the city. This planning process requires that thorough studies of the community be made as a basis for estimating and planning future community needs.

In counties, planning is accomplished through the provisions of the County Zoning Enabling Act, Chapter 358A of the Iowa Code. There is no county planning enabling act. The Attorney General's office has ruled, "Since the law requires that zoning be based upon a comprehensive plan of the county, the right of counties to plan is inferred."

Zoning is usually regarded as a tool of planning. Once a plan has been devised and approved by the community, zoning is applied to regulate and control land-use changes in a manner which would guide the growth into the patterns established by the comprehensive plan.

Regional planning is permitted in Iowa by the most recent planning enabling act, which was adopted by the legislature in 1963 and has enjoyed rather wide use since. In brief, the act permits any combination of local governments to join together in a common planning effort. This includes large cities, satellite suburban communities,

## BACKGROUND OF THE PLANNING PROCESS

isolated towns, and the counties. The act defines in broad terms not only the many types of local governing bodies which can join together in the planning process but also which of the many planning activities can be pursued. The logical area to be embraced in a regional planning effort should include a population center and its sphere of social and economic influence.[3,5] The term *functional economic area* has been applied in Iowa to the regional concept.[2] This area is noted to be approximately coextensive with (1) a labor market or commuting area and (2) a retail trade area for the variety of goods and services usually found in cities of more than 25,000 population.

In Iowa this would involve multicounty areas. While most regional planning commissions in Iowa do not measure up to this basic philosophy, the regional planning act does provide a means of considering larger planning units than just one city, one town, or one county. The act has been used most frequently by a county and its political subdivisions, such as all of the incorporated cities and towns in the county. Adams County was the first to form a regional planning group under the provisions of Chapter 110. In one joint effort four counties in northwest Iowa have embarked upon a regional planning program. The Black Hawk Metropolitan Planning Commission in the Waterloo area is an example of a cooperative effort in a metropolitan region. The regional planning act provides Iowa communities with a comprehensive tool with which to attack flood plain problems. For instance, planning does not have to stop at the corporate line, but can be coordinated throughout the rural and urban reaches of a river. This is particularly useful in coordinating industrial development and suburban fringe developments of a residential nature.

### EXTENT OF AUTHORITY

It should be pointed out that planning commissions or zoning commissions have no direct authority or power; this lies in the hands of such local legislative bodies as the city council or county board of supervisors. These commissions have only the power to recommend that certain action be taken. For planning and zoning decisions of these commissions to become policy or law, they must be approved and legally enacted by the local legislative body. For this reason it is important that planning commissions be thorough and reliable in their studies and recommendations.

Iowa law permits communities to designate the planning commission as the zoning commission. In such cases it is usually entitled the "plan and zoning commission." In most Iowa cities and towns, the zoning commission and the planning commission are the same body.

## COMPREHENSIVE PLANNING

The principles of community planning and zoning may be applied directly to flood plain areas. The most advantageous approach consists of including the flood plain within the total scope and areal extent of the community or regional planning program. Within this concept the general planning process will be reviewed briefly.

There are several phases of the general planning process upon which the comprehensive plan is based. Those elements most frequently discussed are included in the following list:[1,4]

1. Mapping—to provide up-to-date maps for study and planning purposes.
2. Existing land-use survey and analysis.
3. Economic analysis and forecast.
4. Population analysis and forecast.
5. Future land-use plan.
6. Transportation analysis—including primary and secondary streets; public transportation; and other modes of travel, including air and rail.
7. Central business district analysis.
8. Industrial development studies.
9. Urban renewal or area rehabilitation plans.
10. Community facilities review—schools, parks, and other public facilities.
11. Utilities—extensions and improvements.

The availability of assistance through ongoing programs at the state and federal levels has permitted an accelerated rate of planning in recent years. Special attention needs to be directed to flood plain areas in these planning phases.

Comprehensive planning programs can be effectuated through four primary measures: (1) the zoning ordinance, (2) the subdivision regulations, (3) the building code, and (4) the capital improvement program. Other special authorities or programs may be needed in large metropolitan areas for accomplishing programs of urban renewal, rapid mass transportation, and the like. Through these measures and additional programs, local officials and citizens may direct the growth of their community. Annual review, report, and implementation of changes and adjustments are necessary to keep the planning program up to date.

## SUMMARY

Communities through state planning legislation are permitted to plan for and to regulate land uses in flood plains. In Iowa local or regional programs in flood plain planning and zoning must be coordinated with the statutory provisions administered by the Iowa Natural Resources Council. The responsibility for the preparation of plans and local regulatory measures lies with duly organized local planning units. The responsibility for carrying out the plans and enforcing regulations rests primarily on the shoulders of local legislative bodies. Assistance of the state in regard to flood plain activities in Iowa should be of immeasurable value in accomplishing sound programs of flood plain management at the local level.

## REFERENCES

1. Austin, W. B. The Yankton, South Dakota, planning program. *IPA Newsletter* 14, no. 80. Iowa City: Iowa Planning Association. 1966.
2. Fox, Karl A. *Functional economic areas.* An open letter delivered to Economics Workshop on Planning. Iowa State Univ., Ames. Jan. 7, 1965.
3. Friedman, J., and Alonso, W. *Regional development and planning, a reader.* Cambridge: MIT Press. 1964.
4. Goodman, W. I., and Freund, E. C. *Principles and practice of urban planning.* Washington: International City Managers Assoc. 1968.
5. Perloff, H. S., et al. *Regions, resources, and economic growth.* Baltimore: Johns Hopkins Press. 1960.
6. Peterson, C. E. Iowa state laws, policies and programs pertaining to water and related land resources. *Proc. Fourth Annual Water Resources Design Conf.* Ames: Iowa State Univ. Eng. Ext. 1966.

# 7

# LAND-USE PLANNING FOR THE FLOOD PLAIN

### William R. Klatt

ALL OF US RECOGNIZE the pressure of urban growth on flood plain areas and, simultaneously, a strong desire for conservation of our natural resources, including our river valleys. In this section we shall discuss flood plain planning rather than regulation. Optimum utilization of the flood plain for the benefit of all residents of the community can only be achieved by applying sound and realistic planning concepts.

## DEFINITIONS

Two definitions adopted by the American Society of Civil Engineers will be used to illustrate flood concepts.[1] The *flood plain* is the relatively flat area or lowlands adjoining the channel of a river, stream, watercourse, lake, or ocean which has been or may be covered by floodwater. A *flood* is an overflow of lands used or usable by man and not normally covered by water.

WILLIAM R. KLATT is an Urban Planner with Stanley Consultants, Inc., Muscatine, Iowa.

Floods have two characteristics. The inundation is (1) temporary and (2) comes from an adjacent river, stream, lake, estuary, or ocean. Adverse effects include damage from overflow, sediment deposition, sewer backup, backwater in drainage channels, creation of unsanitary conditions, rise of the groundwater table, and other problems. The Mississippi River floods of 1965 and 1966 showed all of eastern Iowa the relentless power and strength of a river in flood. The tremendous effort expended in the United States in flood fighting, cleanup, and rehabilitation illustrates that man periodically pays a price for flood plain occupancy.[7,8]

We should distinguish next between the natural floodway and the restrictive zone, or floodway fringe. The natural *floodway* is the main channel and adjacent overbank areas which convey floodflows with an observable velocity. The *restrictive zone,* or *floodway fringe,* is the flood plain area landward of the natural or designated floodway which still would be inundated by the design flood selected for regulation or control purposes. These slackwater areas, with little or no velocity of flow, are the flood plains most easily and frequently occupied by man. These concepts are illustrated in Figure 7.1.

The essence of the differences between the two zones, in planning terms, lies in the ranges of uses possible in each. The floodway, left

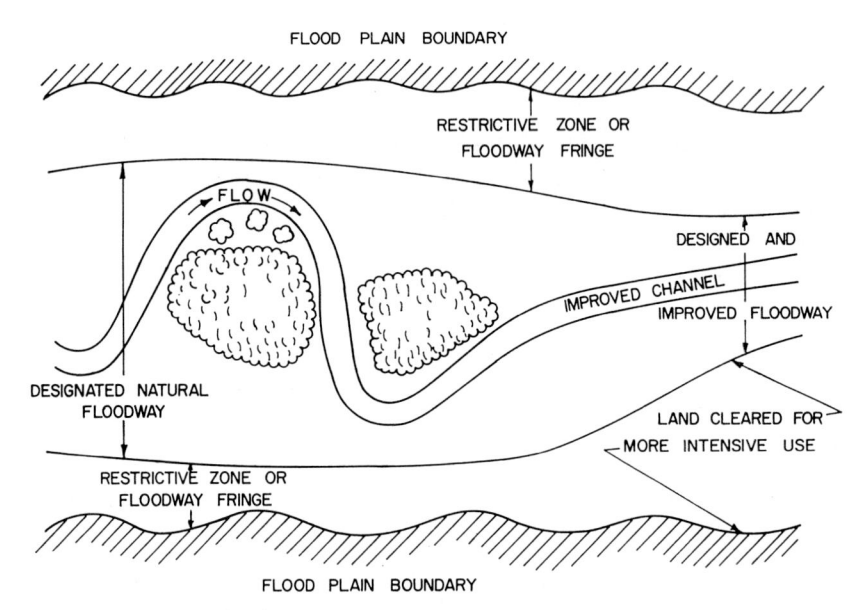

Fig. 7.1. Floodway and restrictive zone or floodway-fringe concepts for the flood plain.

# LAND-USE PLANNING FOR THE FLOOD PLAIN

in its natural state or improved, must remain unobstructed by works of man so that floodflows can be discharged adequately and efficiently. Under certain conditions the restrictive zone can be developed for more intensive urban uses.

## FLOOD PLAIN CHARACTERISTICS AND LAND-USE RELATIONSHIPS

Returning to the planning aspects, we might borrow from a modern idiom and consider "the message and the medium." In our case the message involves land uses based upon needs, or real development potential in a specific situation. We want to know what is essential for community growth and development as opposed to what will happen if development takes place in the absence of planning. The "medium" then becomes the flood plain. A complete understanding of the medium is implicit. We have already entered a phase of planning encompassing entire watersheds, and thus we are forced to evaluate regional as well as local effects.[4] Channel constrictions benefiting one area may have drastic effects on nearby flood plains. Stream straightening may improve land utilization in one reach but be a serious detriment to those downstream. Briefly, physical characteristics of the channel and flood plain determine hydraulic capacity, velocity of flow, and depths of flooding. Within a selected study area the planner must analyze cross-sectional area, shape, roughness, slope of the channel, and slope of the flood plain.

In many cases, the planner must anticipate what changes man is going to make or attempt to make in these hydraulic characteristics. The reason for changing the flood plain configuration is as vital as the effects of the change. All of these may have a bearing on selecting appropriate land uses for flood plain areas. They have a special significance in selecting control elevations for urban development that must be constructed above the predicted flood elevations.

### Uses Under Natural Conditions

If the definition of flood plain is recalled, then the list of land uses acceptable without requiring some sort of countermeasures can be made very brief. In other words, if we are to use the flood plain without materially changing its form, dimensions, and other vital statistics, it is suitable for such low-damage uses as:

1. Conservation and wildlife practices.
2. Agricultural crop production and livestock farming.

3. Open-space uses involving recreation, utilizing low-damage, open type buildings and structures.
4. Transporation elements of a secondary nature, such as low-lying streets which can be closed off periodically, when inundated, without causing complete disruption of the transportation system.

A few natural uses may have been omitted, but generally other land uses or protective measures would change the cross-section area, roughness, or configuration of the flood plain. These in turn would affect the flood carrying capacity and might easily cause a detrimental change in flood conditions upstream. We must not forget that what is done in Muscatine's flood plain may have some effect as far upstream as Rock Island and Davenport. The problems of alleviating such effects will be discussed later in this chapter.

## Planning for More Intensive Urban Uses

The planning consultant frequently is concerned with various types of flood plain situations in urban and regional planning studies. This possibly is due to many problems encountered in the competition for intensive urban land use. Along the Ohio River, for example, several cities rest completely behind floodwalls. Some are 16 feet high, and extreme separation of city and river then becomes a reality. The river and urban area exist under a sort of armed truce. Familiar to our reach of the Mississippi River are earth levees and concrete floodwalls of various heights.[6] Although they are not as massive or striking as the Ohio River floodwalls, the separation is just as permanent and definite. There is little if any esthetic value of a valley location for a community under these circumstances.

Without resorting to flood protection or floodproofing devices, land uses and structures are often placed in the flood plain, based upon predicted flood frequencies and predicted flood heights. Recreation features, river transportation facilities, and some transportation routes may have to occupy flood plain sites to perform their intended functions. Floodproofing techniques, special structural considerations, special site studies, and special handling of utility features are necessary for such installations.[5] Recently designed and under construction is a large bathhouse and beach at a flood control reservoir which is the central water feature of a large state park. The flood control factor cannot be compromised. Therefore, the facility had to be designed for the expected inundation. Flood levels at the beach will be exceeded annually, and the roof of the bathhouse may be covered once in 25 years. Special structural design will permit periodic inundation, but with a minimum of damage.

The Ohio River is subject to rapid rises which are much higher than those experienced on the Mississippi River. In flood plain areas at Cincinnati and Portsmouth, Ohio, great variances in annual flood stages are encountered, with river stages on the order of 35 feet above low water being anticipated each year. Docks and small boat accommodations must be sectionalized for easy removal. Walks and promenades are designed to be cleared of mud and sand with high pressure hoses. All utility connections can be closed off by waterproof fittings during high water.[5]

Today design features can be included that accommodate specific uses despite periodic inundation, due to better information, new technology, and new materials. A more ideal method, overall basin water management, is however a more systematic application which deserves our primary attention today. This method permits more intensive flood plain use with a minimum of land-use interruption by flooding.

## RIVER BASIN WATER MANAGEMENT SYSTEMS

A few outstanding systems of water management exist which embrace virtually the full range of flood plain and flood control measures. The Miami Conservancy District in Ohio was legally organized under a new state statute in 1914, following the catastrophic floods of 1913 in Dayton, Hamilton, and Middletown, Ohio,[4] and provides a high degree of flood protection to millions of dollars of property value in nine counties. Five large dams and reservoir sites store water temporarily during high flows and release it at a controlled rate. Levees and channel improvements have been constructed in the urban concentrations. Channels are owned through rural areas with easements or ownership of flood plain areas. Rainfall is monitored and a computer mathematical model permits the volume and rates of flood runoff to be estimated.[2] Discharges from storm and sanitary sewer outfalls also are calculated.

The entire system is supported financially by the property which is protected. No federal assistance was available at that period of the century, in the 1910's and 1920's. Assessments were, and are today, assigned against each benefited property. Incidentally, the original construction is now debt-free. The technical and organizational aspects of the Miami Conservancy District are impressive, but even greater is its fundamental approach and concept that floodwaters do not disappear or diminish by passing through such flood protection measures as levees and floodwalls. Management must be on a watershed basis. In this case the large reservoirs which remain dry most of the year are an essential element. They remain in agricultural uses or in open space and recreation programs during dry-weather periods. During flood periods the reservoirs store excessive volumes of floodwaters which the downstream channels cannot accommodate. Designers of this early system applied the test of land uses appropriate to the flood plain, relating effect of the flood height, velocity of flow, and flood frequency on existing and prospective land uses in the river basin. An optimum system was achieved for the flood plain lands, balancing desired land use with flood protection measures.

Additional examples can be found today, incorporating the multipurpose water-use approach.[4] The Upper Missouri River basin is representative with the system of main-stem and tributary reservoirs and main channel improvements which has been constructed. These comprehensive river basin systems, when fully coordinated with comprehensive planning of all flood plain uses, would provide a stable reference level for flood plain development. Construction of planned programs at one urban center would not then cause an unplanned and adverse effect at other locations. Additional planning on this basis in the Upper Mississippi River valley should be directed toward a basin management system. This would be much preferred over the chain-reaction problems now created by individual communities attempting to correct or improve their local planning and land-use activities with little or no concern over the effects upon other areas.

## SUMMARY

The flood plain could be completely abandoned insofar as beneficial land use is concerned. This is neither a practical nor an economical answer, considering the existing investment in flood plain development. Aside from abstinence, our present thoughts about planning produce two concepts for use of flood plain lands. The first embraces use of such basin-wide flood protection systems as levees, floodwalls, and other structures, provided their effects upon the watershed are calculated and assessed and that the social benefits outweigh the costs and other negative factors. This would be a policy of continuous land-use adjustment backed by an information system encompassing the entire river basin with its tributary watersheds.

# LAND-USE PLANNING FOR THE FLOOD PLAIN

The second concept or alternative introduces a kind of rotating supply or use of flood plain lands. Thus, land within the flood plain could be built upon, raised in elevation, or surrounded by individual or local protection works if it were economically feasible and could be replaced in kind and at an appropriate place in the river basin. This concept would provide for continuity in the amount of land devoted to specific desired land uses.

Both systems have added significance, as our recreation and transportation demands increase.[3] The conservation-adjustment concept offers the planner opportunities for a time-sequence of land uses in the flood plain. Land reserved during the 1960's as public open space or for crop production may be mined for gravel or limestone during the 1970's and returned as a lake and recreation area by 1980.

Wise management of natural resources as dynamic and powerful as our rivers and streams calls for consideration of all lands affected. The Mississippi Valley has seen the effects of higher and higher flood stages. A systems approach to flood plain development including replacement programs to balance flood damages and loss of land to urban uses is needed as part of a water control system equal to the urban growth and climatological consequences which we face.

## REFERENCES

1. American Society of Civil Engineers. *Guide for the development of flood plain regulations*. Progress rept., Task Force on Flood Plain Regulations. *Proc. Am. Soc. Civil Eng., Hydraulics Div.* 88, no. HY5, Paper 3264 (Sept. 1962).
2. Doyle, T. J., and Thomas, R. F. *Development and verification of a flood routing model of the Miami River basin*. Restudy of the official plan, pt. 1. Dayton: Miami Conservancy Dist. 1964.
3. Levine, L. Land conservation in metropolitan areas. *J. Am. Inst. of Planners* 30, no. 3 (Aug. 1964).
4. Linsley, R. K., and Franzini, J. B. *Water resources engineering*. New York: McGraw-Hill. 1964.
5. Sheaffer, J. R. *Introduction to flood proofing, an outline of principles and methods*. Chicago: Univ. of Chicago Press. 1967.
6. U.S. House of Representatives, Committee on Public Works. *Mississippi River, urban areas from Hampton, Illinois, to Cassville, Wisconsin*. House Document 450, 87th Cong., 2nd sess. 1962.
7. White, G. F., et al. *Changes in urban occupancy of flood plains in the United States*. Dept. of Geog. Res. Paper 57. Chicago: Univ. of Chicago Press. 1958.
8. White, G. F., ed. *Papers on flood problems*. Dept. of Geog. Res. Paper 70. Chicago: Univ. of Chicago Press. 1961.

# 8

# REGIONAL CONSIDERATION
# OF FLOOD PLAIN PLANNING

William S. Luhman

THE BI-STATE Metropolitan Planning Area in Scott County, Iowa, and Rock Island County, Illinois, is a suitable example of the necessity of regional planning for the sound management of flood plain areas. The two-county, two-state area is bisected by 62 miles of the 2,470-mile-long Mississippi River. Within the Mississippi River flood plain in this planning area are 14 cities and towns, 11 townships (Illinois), 17 school districts, 7 fire districts, 2 drainage districts, 1 forest preserve district, 1 airport authority, 2 bridge commissions, 2 levee commissions, 2 counties, and 2 states. With the federal government this totals 61 agencies of government.

The total population of the two-county area was over 270,000 in 1960.[5] The area experienced a 15 percent increase in population between 1950 and 1960, and the growth trend has continued in the 1960's. The major cities within this region, Davenport and Bettendorf, Iowa, and Rock Island and Moline, Illinois, have substantial portions of their urban areas located on the flood plain.[9] The two

WILLIAM S. LUHMAN is Executive Director of the Bi-State Metropolitan Planning Commission for Scott County, Iowa, and Rock Island County, Illinois.

93

counties (with Henry County, Illinois, added) constitute one of the standard metropolitan statistical areas for the U.S. Bureau of the Census, one of 224 in the United States. This illustrates the regional importance of these metropolitan counties.

In this chapter attention will be focused upon the development of effective regional planning programs which relate to flood plain management. Among the concepts which concern regional planners are the federal influence, cooperative multi-use community facilities, regional water-use problems, organization structures for planning and programming purposes, central authority needs, and coordination techniques.

## RECENT TRENDS TOWARD REGIONAL PLANNING

The single governmental unit approach to land-use planning of flood plains is becoming recognized as an increasingly more futile effort.[3,4] The tremendous growth that the urban areas of our nation are experiencing has been the result not only of sizable population growth but also of the increasing concentration of this population around our industrial centers of employment. Accompanying these trends has been a general increase in the national and regional economy. This has resulted both in increased income for consumer goods and services and in a swift advance in technology to support development of new products. This train of events has significantly altered urban living patterns.

One major result, among many others, was the massive production of automobiles which provided almost complete mobility to the population lying within and between these urban concentrations. Today this mobility seems to provide a limitless expanse for urban growth. This growth or "urban sprawl" disregards the seemingly artificial political boundaries of cities, counties, drainage districts, and the states. As this sprawl envelopes the flood plains of major rivers and small tributary valleys and even the smaller ravines, it compounds our urban problems with those of poor drainage and flooding.

### Federal Interest in Regional Planning

Our federal system of government preserves many rights for the states, including the formation and control of local government entities. The result is a pluralistic local government system that today requires regional cooperative planning for the continued growth and development of these urban regions. The typical problem within a given urban region as far as local governments are concerned is that there may be a number of counties, cities, and towns (and not infrequently a number of special-purpose districts) superimposed upon the others, each with some authority to regulate land uses or construct various public facilities within a given urban area. They may also proceed to carry on their respective functions and capital-improvement programs independent of one another. There has been no real coordination or unified plan for these agencies other than a recent edict which permits counties to combine certain functions.[2]

Federal policy has reflected concern about this problem and the need for coordinating regional development decisions in urban areas. The Federal Aid Highway Act of 1962 required that in urbanized areas of 50,000 or more population a continuing cooperative transportation planning process must be developed among the local governments and the respective state highway departments.[6] Any federally financed highway construction must be based upon a comprehensive regional plan developed through this planning process. The federal government does not intend to design the land-use plan for the respective urban areas, but it is requiring that the decision-making organizations of those areas organize themselves to make decisions on how the area as a total region will develop.

The Federal Aid Highway Act provided a July, 1965, deadline for the completion of these comprehensive plans. While this generally has not been met in most of the metropolitan areas, it has served as a means for strongly promoting the organization of either regional councils of government or regional planning commissions to carry out this planning process.

Closely following the requirements of the Federal Aid Highway Act were the metropolitan planning requirements of the U.S. Department of Housing and Urban Development (HUD) and the regional planning requirements of the Farmers Home Administration.[8] Here again, the basic intent is to encourage (if not to require) local communities within urban regions or in rural county areas to join in cooperatively developing and implementing a comprehensive plan for the growth and development of the particular region.

The latest reflection of federal policy regarding regional planning has been expressed in terms of the Demonstration Cities and Metropolitan Development Act of 1965.[7] This act provides for the designation of local area-wide review agencies which have responsibilities to review applications for federal grants for the following facilities:

1. Open-space land and recreational facilities.
2. Hospitals.
3. Airports.
4. Libraries.
5. Water supply and distribution systems.
6. Sewerage facilities and waste-treatment facilities.
7. Highways.
8. Other transportation facilities.
9. Water development and land conservation.
10. HUD and USDA planning grants.

These programs relate to a number of facilities that can assist in making a sound flood plain management program more effective.

Thus urban areas of 50,000 or more people are confronted today with the requirement to develop a regional plan cooperatively. In addition, they are required to submit their applications for federal grants, for costsharing of various capital facilities, to a designated areawide review agency to ensure that such facilities are consistent with a metropolitan plan.

### Additional Cooperative Ventures

Recognition should also be given to a new development in regional planning and cooperation. This is the great increase in the organization of councils of government throughout the country. Increasingly, elected officials are giving recognition to the necessity of individual local governments joining together in solving their related and frequently interdependent urban problems. The involvement of elected officials in these activities emphasizes their recognition of the serious problem of regional growth, and perhaps of their determination to implement on a regional level the area-wide plans that are prepared.

## SEQUENTIAL PLANNING STEPS FOR REGIONAL LAND USE

In the light of the multiplicity of governmental agencies having jurisdiction in matters related to flood plains, we need to increase regional effectiveness in solving flood plain problems. Achieving programs for sound management rests primarily with those responsible for comprehensive regional planning. Assuming the existence of a responsible agency capable of sound land-use planning, the steps in this process can be defined.

**Goals and Objectives**

Area-wide goals and objectives for policies relating to the management of flood plain areas should be evolved. These might be developed from a broadly based representative group of citizens serving in an advisory capacity to the regional planning commission. Such a group can be helpful in bringing to the commission various viewpoints of the community-at-large and taking from it a better understanding about the problems of the region and the purposes and functions of the regional planning commission. It is hoped the result will be a stronger region-wide commitment to planning and a better understanding of the desirability of a good flood plain management program.

**Cooperative Regional Planning**

Through general comprehensive planning inventories and analyses of the regional agency, the existing problems, plans, and potentials for sound flood plain management may be identified. The cooperation and participation of all responsible local governments as well as the many functional agencies of the state and federal governments are essential for the appropriate consideration of all significant interests. The general planning inventories and analyses with respect to population, economic base, land use, physical resources, public facilities, and transportation facilities must be considered in the light of the state and federal governments' many functional agencies planning for any of the following beneficial uses of water:[10]

1. Domestic, municipal, and industrial water supply.
2. Irrigation.
3. Water quality control.
4. Navigation.
5. Energy production, including hydroelectric, thermal-electric, and nuclear powerplants (now the concern is more with thermal-electric and nuclear power, with the accompanying requirement

for makeup water and for cooling of the circulating water).
6. Flood plain management, including flood control and regulation to reduce losses.
7. Land and beach stabilization.
8. Drainage including salinity control.
9. Outdoor recreation.
10. Fish and wildlife.
11. Other purposes contributing to economic growth and development.

The relationship of water resources to land-use planning means it is essential to maintain an effective working relationship at all professional levels. Coordination must be achieved between the regional planner and the engineer, the economist, the public health official, and the many other specialists whose knowledge and training are required for adequate appraisal of the development problems of flood plain areas.

**Regional Land-Use Plan**

The preparation of a regional plan for the management of flood plain areas is actually an integral but primary consideration of a total land-use plan. Consideration must be given to the goals and objectives of the planning efforts and to the findings and conclusions drawn from the inventory and analysis of the region. Alternative consideration in selection of areas for future growth need to be included.

In regions with many local governments and various state and federal agencies, it is desirable to proceed from selection of the very general long-range alternative-concept plan to the development of the somewhat more detailed land-use plan and accompanying individual elements, many of which will relate to flood plain areas. Through the development of alternative-concept plans, the factors affecting desirable urban growth can begin to be synthesized into alternative concepts or schemes for guiding regional growth. These alternative concepts can be evaluated by the affected governments, and the decision-making process toward a regional plan can begin.

**Effectuating a Regional Plan**

The development of a regional plan and its several elements is not an end in itself.[3] The real concern is how to transfer the plans

## REGIONAL CONSIDERATION OF PLANNING

to accomplishments. This includes the preparation of policies, programs, and projects within the formidable framework of our pluralistic system of local governments. This phase must be considered during the preparation of the regional plan if it is going to have any degree of acceptance and assurance of implementation by the many affected jurisdictions. The local powers of zoning control, subdivision regulations, mapped streets act, capital improvements, and the policies and programs of the many state and federal functional agencies should be fully integrated to achieve the most effective flood plain management program possible.

While the regional planning commission generally does not exercise any of the power and authority over local land-use control or capital-improvement construction programs, its close attention to the problems and potentials of this phase of the program on a continuing basis is essential if the regional plan is to be implemented to any degree. The regional planning commission's greatest resource in implementation is its potential role as a sort of "regional conscience" for the long-range good of the area.

## COORDINATION PROBLEMS

The regional planner must assume the responsibility for ensuring that the general interests of the entire region will be considered by local agencies having specific functional responsibilities. It is also his role to evolve effective and reliable mechanisms for joint and continuous planning as a part of the comprehensive regional planning effort. Care must be given to develop the kind of definitive and useful plans from which proposals for various projects can be prepared. To be effective, regional planning must fit into the process by which the local government and the various functional agencies make both short- and long-range decisions relating to the management of flood plains.

A major limitation to the effectiveness of this planning effort is the lack of a central authority to implement these plans. The pluralistic local government maze and the federal system provide major obstacles to any ease of implementation.

### The Bi-State Commission

The Bi-State Metropolitan Planning Commission's area is confronted with many of the problems of regional planning—indeed, the problem of regional planning considerations for flood plain management is of special significance.[1] This two-state, two-county area not only is bisected by the Mississippi River but also is greatly affected by two others. The Wapsipinicon River in Scott County, Iowa, borders Clinton County to the north of the planning area; and the Rock River in Rock Island County, Illinois, borders Henry County to the south. Not only are two states and two counties included but four additional adjacent counties outside the planning area are also involved in the regional flood plain picture. These are illustrated in Figure 8.1 which shows the many local communities in the regional planning area.

The organization for the planning process in this two-county area is supervised by the Bi-State Metropolitan Planning Commission, consisting of 22 members—11 from each county or portion of the two-state area. Four of 11 from each state are elected officials, 1 being

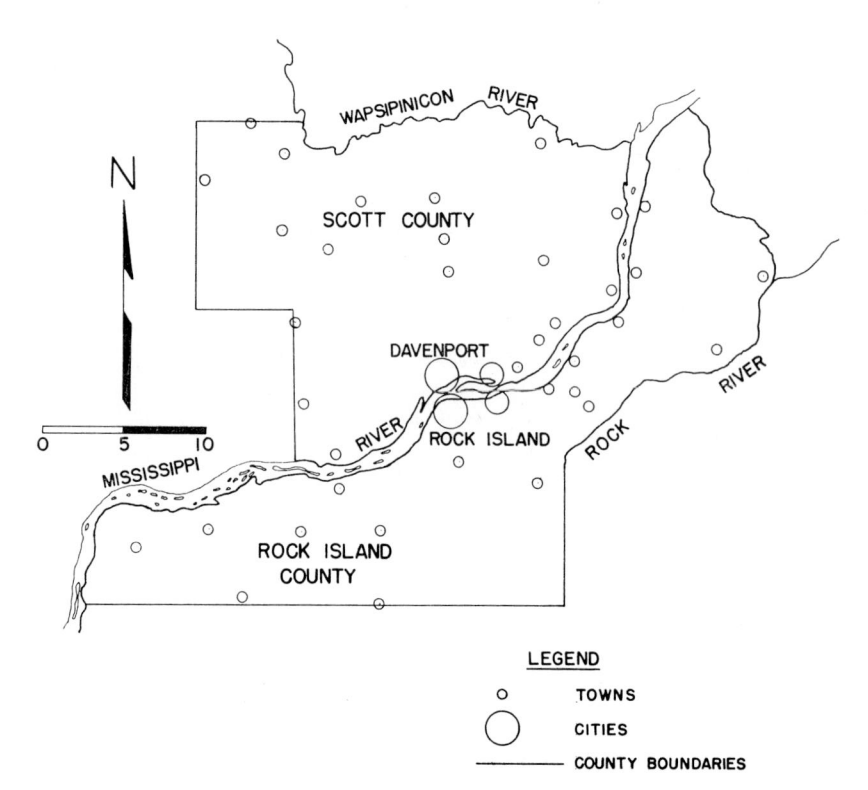

Fig. 8.1. The two-county area included in the Bi-State Metropolitan Planning Area, Scott County, Iowa, and Rock Island County, Illinois.

the chairman of the respective county board of supervisors and 3 being the mayors of the three largest contiguous cities on each side of the river. The remaining 7 from each county are citizen members appointed by the local governing bodies according to the respective state's laws. In addition to the planning commission, there are a series of technical committees consisting of city planners, city engineers, park and recreation directors, and directors of departments of public works for the six cities and two counties. These committees function for transportation, outdoor recreation and open space, and sanitary sewer and water systems. Various state and federal officials—that is, Corps of Engineers, U.S. Army Weapons Command, U.S. Bureau of Public Roads, state highway departments, conservation commissions, and state development commissions—serve as advisory members to these technical committees. In addition to the technical committees, there is a 140-member citizens' advisory committee which includes at least one elected official from each town, village, and other special-district local government within the two-county area as well as citizen representation from civic groups having an interest in the development of the region.

When the flood plain aspects of this regional plan are completed, they will represent the collective thinking of the two-county area and the functional agencies of the state and federal government. It will be a guide for the management and development decisions of these agencies. It is hoped that it will be a means of maintaining a continuing regional planning effort and will achieve the program coordination that is necessary to respond to the dynamics of the region.

## CONCLUSIONS

We should understand clearly the major limitations of this regional approach. We must recognize first what has already been pointed out—the great limitation of planning implementation with a pluralistic system of federal, state, and local governments and a regional planning function that is advisory only.

In addition, the example of the Bi-State Metropolitan Planning Commission planning area indicates the futility of regional planning without similar programs in operation along the length of the river, based upon a total river basin program. The real contribution the regional approach makes is to begin coordinating the many local jurisdictions within each of the respective regions along the river.

River basin planning could be coordinated in this bi-state area of the Upper Mississippi River with one regional interstate agency instead of 62 agencies involved in the 62 miles of the 2,470-mile river, a present average of one agency per mile. The need for continued regional efforts is obvious.

**REFERENCES**

1. Bi-State Metropolitan Planning Commission. *A summary report, comprehensive plan, bistate metropolitan planning area.* Rock Island. 1968.
2. Iowa Code. Ch. 332. 1966.
3. Mayer, Albert. *The urgent future: People, housing, city, region.* New York: McGraw-Hill. 1967.
4. Spieloogel, Samuel. *A selected bibliography on city and regional planning.* Washington: Scarecrow Press. 1951.
5. U.S. Bureau of the Census. *County and city data book, 1967.* Washington: USGPO. 1967.
6. U.S. Code. Sec. 134, Title 23. 1965.
7. U.S. Congress. *Demonstration cities and metropolitan development act of 1966.* PL 89-754. 89th Cong., 2nd sess. 1966.
8. U.S. Department of Housing and Urban Development. *Programs of HUD.* HUD IP-36 Washington: USGPO. 1967.
9. U.S. House of Representatives, Committee on Public Works. *Mississippi River, urban areas from Hampton, Illinois, to Cassville, Wisconsin.* House Document 450, 87th Cong., 2nd sess., 1962.
10. U.S. Senate, Select Committee on National Water Resources. *Water resource activities in the United States.* 86th Cong., 1st sess., 1959 and 86th Cong., 2nd sess., 1960. See Index for Committee Prints 1–32 (Prints 1–32 published in the complete series) 86th Cong., 2nd sess., 1960.

# 9

# COORDINATION OF URBAN PLANNING AND FLOOD PLAIN DEVELOPMENT

### Eugene O. Johnson

THROUGHOUT HISTORY men have built cities within flood plains and then attempted to control and protect them from flooding by constructing floodwalls, levees, and other flood control devices. For the most part these are expensive and are never completely effective. As rapid urbanization creates greater amounts of runoff and the construction of additional levees and floodwalls creates continually higher flood stages, the problems are compounded. More recently, management and controlled use of flood plains have been introduced as being more optimum methods of reducing flood damages.[10]

Both approaches and their advantages and disadvantages are discussed elsewhere in this book. Rather than belabor these points, it appears appropriate here to review a "middle of the road" approach developed at Davenport, Iowa, as a solution to the problem of flood plain occupancy and attempts to control new construction. The efforts to achieve a solution were not solely those of the city of Davenport, the Iowa Natural Resources Council (INRC), or the U.S. Army Corps

EUGENE O. JOHNSON is Director of the Plan and Zoning Commission, Davenport, Iowa.

of Engineers, nor were they solely those of the engineers or planners; but all of these agencies and professions were working together in a coordinated effort.

While this effort has resulted in a program that looks good on paper, it has not been tested in practice. Efforts have been unsuccessful in getting the program at Davenport adopted because of a lack of coordination and/or understanding at certain levels. Several new and relevant concepts evolved from the initial attempt to implement a regulatory program, and they deserve presentation in this discussion of flood plain management problems.

## THE DUCK CREEK PROBLEM

Davenport is bisected in approximately its geographical center by a usually placid, at times almost intermittent, stream called Duck Creek.[9] The majority of the incorporated area of the city lies within the Duck Creek watershed, as shown in Figure 9.1. This creek, like most other creeks and watercourses which are directly tributary to the Mississippi River, sometimes goes on the rampage.[5,6,11] Although urban development now has spread beyond Duck Creek, until a few years ago its flood plain was almost entirely devoted to agricultural or open-space uses. The flood damage potential was therefore at a minimum. Recently, with land becoming more at a premium, developers

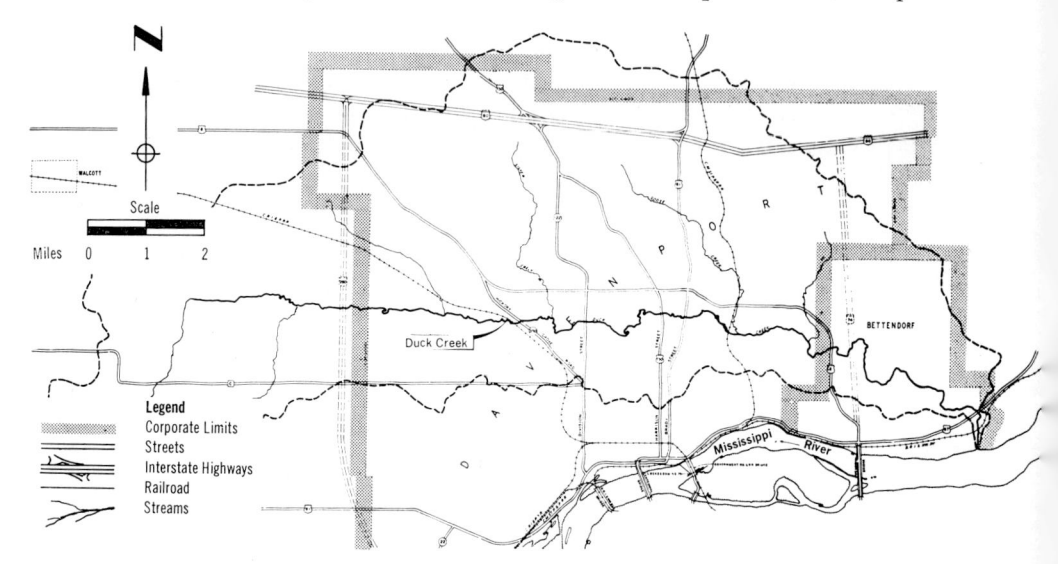

Fig. 9.1. Duck Creek at Davenport, Iowa.

began to subdivide land that previously appeared undesirable. This increasing urbanization and encroachment into the flood plain made imperative the need for an effective method of control.

The first record we have of any recognition of a flood problem on Duck Creek (or of attempts at a solution) is a newspaper article, dated December 31, 1918. Referring in all probability to the first engineering study prepared for Duck Creek, it shows a picture of the flooded Duck Creek flood plain. The article is captioned "Davenport's Biggest Improvement Project This Year Will Be To Prevent Repetition of Scenes Such as This" and goes on to suggest that the best solution to the problem of flooding along the Duck Creek valley was the creation of a park throughout the length of the creek for recreation purposes and to provide right-of-way for sewer facilities. This park was to be 600 feet wide throughout its length. The land could be obtained for very little cost at that time and the city of Davenport could proceed to install sewers in the purchased land.

During the early 1920's the city initiated this project, beginning with construction of a district sewer that ran along Duck Creek. The 600-foot width of park area proposed by the original study was not obtained, however. Only small rights-of-way were purchased for the sewer project itself. Although this trunk sewer project has been beneficial in encouraging development of the northern part of Davenport, it has done very little to control or manage the Duck Creek flood plain. In fact, the only land purchased as part of this proposed 600-foot strip was a 50-foot right-of-way for a proposed parkway that would run parallel with Duck Creek for the entire length of the city. This purchase was made in the early 1920's.

The original idea of constructing streets along the creek, with an improved channel within the 600-foot width, has been continued through the years until as late as 1960. At this time the city plan commission, with the assistance of the city engineer, was developing plans for the construction of an improved floodway and pilot channel, with parallel roads along either side. The proposed flood control channel was to be 400 feet wide and would be constructed through parks and farms as well as developed areas. Fortunately, two things happened before it could be adopted as policy.

First, Black Hawk Creek, located south of the Duck Creek watershed, began to flood frequently. A channel similar to the one proposed for Duck Creek had been constructed here and was adequate until sediment deposition and undergrowth clogged it, after which it could no longer accommodate the flood discharges during peak runoff periods. The result was the frequent flooding of surrounding residential developments. It was necessary for the city to expend large sums to clean the channel, raise a deflecting levee, and ensure flood protection for residents in the immediate area. This experience created misgivings about the physical and economic feasibility of constructing and maintaining an open channel as a flood protection measure.

Second, the federal government was requested to participate in the cost of the construction of the channel and necessary levees along Duck Creek. After a feasibility study by the Corps of Engineers, the city was informed that the federal government would not participate because the cost of the improvements could not be justified by the benefits achieved. In other words, there was not sufficient urban development existing to justify flood protection measures economically.

## EARLY ATTEMPTS TO REGULATE THE FLOOD PLAIN

The next approach was to obtain strict control of development within the flood plain. The only real information concerning previous floods on Duck Creek was for one which occurred in 1949, the greatest of recent record. Floods of lesser magnitude occurred in 1959, 1960, and 1963.[9] According to the Corps of Engineers report, a flood of the magnitude of 1949 could be expected once every ten years. Using the limits of inundation from this flood as a guide, an ordinance was enacted to protect and control future development.[2]

This ordinance, which is still in effect but difficult to enforce because of inaccurate original information, provided that "no building permit should be granted by the Building Official for the construction or enlarging of a building for residential or commercial purposes within the flood plain areas, as shown on the 'Official Flood Plain Map,' without first having the approval of the City Engineer and the City Health Official"; further, "that the elevation of any cellar floor or the lowest sanitary or storm sewer lateral at the foundation wall shall be at a higher level than the flood level elevations as established by the 'Official Flood Plain Map'."

In conjunction with this ordinance, the city of Davenport began to take advantage of the assistance offered by the INRC. This state agency reviews developments in urban areas that lie in the flood plain of streams which have a drainage area exceeding two square miles.[3,8] The services of the INRC have been of extreme value, since the city has been able to utilize their technical assistance in fields where qualified local personnel were not readily available. However, this arrangement has not been entirely satisfactory.

# URBAN PLANNING AND FLOOD PLAIN DEVELOPMENT

The coordination procedure between the developer, the city, and the state regulatory agency has proven time consuming for the local developer and has created difficulties because of lack of communication. The developer's engineer has been required to submit plans, have them reviewed, and often revised before receiving final approval. This process may take several weeks or even months. Another problem is the responsibility of ensuring that construction is completed according to the approved plans. The INRC and its staff, because they make the technical review and approval, are best qualified to inspect the work; however, they do not have resident personnel in each city to accomplish this. Further, final enforcement of any subdivision regulations and building improvements lies with the city, since city officials issue the building permits. In the past the city has not individually checked these projects to determine if they comply with the minimum requirements established. The result has been in some instances that construction has taken place not in conformance with approved plans. Once this occurs, it is difficult to achieve corrective measures. This situation has been improved by requiring that a certificate signed by a registered professional engineer be filed with each building permit application. This certificate verifies that the plans approved and the order issued by the INRC have been complied with.

After almost 45 years of studies with various proposals, the results were: one fairly accurately recorded flood, one difficult to enforce ordinance, one technically correct but often cumbersome state review, and about as many proposals and solutions as people and agencies studying the problem.

## FLOOD PLAIN INFORMATION STUDIES

At this point, with the encouragement of the INRC, the Davenport City Council requested the Corps of Engineers to conduct a new study of the Duck Creek valley. The result was a flood plain report completed in 1965[9] which made recommendations for action to be taken by the city of Davenport. It also accurately pinpointed for the first time the limits of the 10-, 50-, and 100-year floods and the standard project flood that could be expected in the Duck Creek valley. At this stage the coordination effort began in earnest. The first proposal by the city plan commission was to keep all the land within the limits of the flood having a 100-year frequency in an open-space use; a use that would suffer no adverse effect if flooded.

The city of Davenport, the park board, the school board, and other public and semipublic agencies had in their control approximately 200 acres in the Duck Creek flood plain. Most of this land was located in several large tracts, with the balance in the small strips which had been purchased in the 1920's for the proposed street and sewer projects. With this much land already controlled, it was hoped that a continuous park or open-space use could be developed that would incorporate all the land within the flood plain, as designated by the Corps report. This met with immediate opposition from many of the landowners and developers in the area because of the amount of land that would be taken out of active productivity and removed from a potential intensive urban use of high value.

There now existed two entirely opposite approaches for solving the problem of flooding on Duck Creek. One was the creation of a pilot channel and improved floodway, which was extremely expensive and would cause problems in maintenance later. The other was the reservation of the area in open-space uses, which was considered to be a waste of land that could be reclaimed for more intensive urban use. At that point the assistance of the INRC was solicited. With the technical information available from the Corps of Engineers report and the combined efforts of the INRC and the city of Davenport, utilizing their engineers, planners, and attorneys, a program was developed.[7] The program basically was to confine a design flood within the area that would normally be inundated by a flood of 10-year frequency. The INRC, using the Corps of Engineers report, determined what land would be inundated by a 10-year flood and the area within these limits that was necessary to convey the design flood.

Since the design flood was considerably larger than the 10-year flood and would inundate greater portions of the flood plain, it was necessary to determine water surface elevations and the water surface profile the design flood would create. In determining these elevations, it was assumed that all land within the 10-year flood limits not needed to convey the flood, as well as all that land outside the 10-year limits which would be inundated by the design flood, would be filled to a safe elevation. Using these criteria, landfill elevations were established that would place any future structures above the control elevations.

## A MODERN FLOOD PLAIN ZONING ORDINANCE

The method proposed to enforce these regulations was flood plain zoning.[1,10] A flood plain district was designated in the zoning ordinance of the city of Davenport when it was revised in 1964; however,

## URBAN PLANNING AND FLOOD PLAIN DEVELOPMENT

no land had ever been placed in this classification. This flood plain district permitted only open-space uses such as agriculture, forests, nurseries, parks, and golf courses. Dwellings, schools, churches, and other structures were not permitted. The ordinance was amended to create two zones or districts for the flood plain.[4,7]

The first was a flood channel district, proposed to include the area necessary to convey flood discharges. The only uses permitted within this area would be the open-space uses previously allowed in the flood plain district. The second would be the flood plain district, which would include the area between the limits of the flood channel district and the standard project line as determined by the Corps of Engineers and presented in their report. This would be a "floating" district superimposed over the underlying zoning. Construction would be allowed within this district in accordance with the underlying zoning as long as the site was filled to the minimum elevation determined by the INRC. Any landowner choosing to develop within the valley would not be allowed to build within the flood channel district, but could build within the flood plain district as long as he filled the land to the established elevation prior to construction of any buildings.

It was proposed that the program be adopted for the entire reach of Duck Creek by both the city of Davenport and the INRC. Administration of the program would be in the hands of the local governing body. This would lift a real burden from developers in the area, since they would no longer have to submit plans to the INRC each time they desired to develop and build in the vicinity of Duck Creek. The proposed procedure would allow the developer to check with the city building officials, where he could obtain data on the limits of permitted encroachment and the minimum elevations required prior to construction. As long as he met these requirements he could proceed with development. If the developer wished to exceed the encroachment limits adopted jointly by the state regulatory agency and the city, he would then be required to apply to the city of Davenport for a zoning change and also to the INRC to determine what effect the changes would have on the overall development. The state agency would then determine what countermeasures, if any, would be necessary. Enforcement and control would rest entirely with the local agency, with a review of any changes by both local and state agencies.

In conjunction with these coordination efforts, the city of Davenport applied to the Department of Housing and Urban Development for a matching-funds open-space grant to purchase an additional 125 acres of land along the creek. The lands selected for purchase were entirely within the flood channel district and would be used as parkway connections between existing city-owned open-space areas.

## FAILURE THROUGH LACK OF PUBLIC ACCEPTANCE

This approach to handling the Duck Creek flood plain problems was legally sound and appeared to meet all the requirements of the planners and engineers, the state agency, and the city of Davenport and was in conformance with the Corps of Engineers report. The proposal was presented to the Davenport City Council by representatives of the Davenport City Plan Commission and the INRC, and a series of public hearings were held in February and March of 1967.

At this time it was apparent that all the coordination between the different agencies and the different professions had fallen short because of a lack of public acceptance and coordination with the political powers. The basic problem of nonconforming uses had been overlooked. These were existing uses which now were illegal because of the zoning that was proposed within the flooded area. The problem was not as severe within the flood channel district as within the flood plain district, that area landward of the conveyance channel. Within this area there were many residential uses which would now be nonconforming and would become ineligible for loans from financial institutions when or if they were sold in the future.

Representatives of local financial institutions testified that they would not provide financing for property zoned in the flood plain category. They added also that if the area were not zoned flood plain they would have no hesitation in financing residential homes. The answer to this was that when flooding occurred the structures would get just as wet whether it was zoned flood plain or not. The people living within these areas by their own admission were not afraid of being flooded, they were merely afraid of what would happen when they tried to sell their property. As a result of public pressure brought about by property owners within the area, the attempt to have the regulation program adopted by the city council failed.

## CONCLUSIONS

After over 40 years of groping in the dark, with the various agencies and professions going their separate ways, efforts to develop a workable flood plain management program have finally been coordinated at Davenport. The coordination, however, is still short of that needed for adoption of the program. The time and energy have not been wasted, however, and a workable solution to the problem of the nonconforming uses presently is being sought. Perhaps in the near future another attempt can be made to have the program adopted. In the meantime, the INRC is utilizing the information gathered and is approving each individual project according to the overall comprehensive plan.

Without coordination the development of adequate solutions for flood plain management and flood control is extremely difficult to achieve. Without continual coordination in program adoption and implementation, a tremendous amount of effort by the many professional people and planning agencies involved in such an undertaking can prove fruitless. Coordination efforts definitely need to involve local residents early in the development of local programs. In addition, initial familiarization of political representatives in city government with the overall program must be accomplished. If these procedures are followed carefully, implementation of flood plain management programs may be accomplished with greater public support.

**REFERENCES**

1. American Society of Civil Engineers. Guide for the development of flood plain regulations. Progress Rept., Task Force on Flood Plain Regulations. *Proc. Am. Soc. Civil Eng., Hydraulics Div.* 88, no. HY5, Paper 3264 (Sept. 1962).
2. Davenport, City of. Zoning Ordinance 122. 1964.
3. Iowa Code. Ch. 455A. 1966.
4. Iowa Natural Resources Council. *A study of flood problems and flood plain regulation. Iowa River and local tributaries at Iowa City, Iowa.* Mimeo. Des Moines. 1960.
5. ———. *An inventory of water resources and water problems, Iowa–Cedar river basin, Iowa.* Bull. 3. Des Moines. 1955.
6. ———. *An inventory of water resources and water problems, northeastern Iowa river basins.* Bull. 7. Des Moines. 1958.
7. ———. *Engineering report on Davenport's proposed Duck Creek development plan.* Mimeo. Des Moines. 1967.
8. ———. *Procedural guide.* Mimeo. Des Moines. 1961.
9. U.S. Army Corps of Engineers. *Flood plain information report, Duck Creek, Scott County, Iowa.* Rock Island: U.S. Army Eng. Dist. 1965.
10. U.S. House of Representatives, Committee on Public Works. *A unified national program for managing flood losses.* Rept. of the Task Force on Federal Flood Control Policy. House Document 465, 89th Cong., 2nd sess. 1966.
11. ———. *Mississippi River, urban areas from Hampton, Illinois, to Cassville, Wisconsin.* House Document 450, 87th Cong., 2nd sess. 1962.

# 10

# FEDERAL PROGRAMS FOR URBAN PLANNING AND FINANCING ASSISTANCE IN FLOOD PLAIN AREAS

Phillip L. Larson

THE INVOLVEMENT of the federal government in programs covering the use and development of flood plain areas is best exemplified by the President's Executive Order No. 11,296 issued in August, 1966. This order is entitled "Evaluation of Flood Hazards in Locating Federally Owned or Financed Buildings, Roads, and Other Facilities and in Disposing of Federal Lands and Properties."[1]

Two excerpts from the executive order are:

> The heads of executive agencies shall provide leadership in encouraging a broad and unified effort to prevent uneconomic uses and development of the nation's flood plains; and in particular to lessen the risk of flood losses in connection with federal lands and installations and federally financed or supported improvements.

and

> All executive agencies responsible for programs which entail land-use planning shall take flood hazards into account when evaluating plans

PHILLIP L. LARSON is an Urban Planner, Department of Housing and Urban Development, Renewal Assistance Office, Chicago, Illinois.

and shall encourage land uses appropriate to the degree of the hazard involved.

There are, of course, many federal agencies with programs that involve land-use planning and development as well as hydrology and other aspects of water resources. This discussion will cover only the programs of the Department of Housing and Urban Development (HUD).

## PURPOSE OF FEDERAL PROGRAMS

The programs of HUD generally are characterized by those that aid planning and development programs being undertaken by state, regional, and local governmental agencies and private enterprise.[2] They are intended essentially to foster good planning and development for the orderly and economical growth of our urban areas. The planning and uses of flood plain lands in (or in the path of) urbanized areas are therefore important elements of these programs.

Secretary Weaver issued a departmental order in May, 1967, entitled "Statement on Policy Implementing the President's Executive Order No. 11,296." The Secretary directed:

> . . . no project, facility, dwelling unit, or area plan for which federal assistance is requested under any of the programs administered by the Department should be so located as to be unduly exposed to potential flood hazard, except where social and economic gains to be derived from such location clearly outweigh the objectives outlined in Executive Order No. 11,296.
> 
> In all cases where there is any indication of a possible flood hazard, in the area where construction will be carried out, hydrologic conditions shall be thoroughly investigated and a supporting technical opinion shall be required. In addition, plans for proposed land usage must be in compliance with applicable federal flood hazard standards before Departmental assistance may be approved.
> 
> Accordingly, engineering review under the respective federal assistance programs of this Department shall, where appropriate, include a thorough consideration of drainage conditions on all sites, including storm or flood water entering and leaving the site and possible ground water problems. Design details should be directed or redirected toward conservation of the natural environment through effective control of storm water. Land use should be appropriate in view of the degree of hazard involved and should eliminate or minimize the need for any future flood protection construction.
> 
> Regional Office staff engineers and District Office underwriters shall study such flood plain data as may be appropriate for review of project plans. If the need for flood control facilities is apparent or if it is questionable, available information of flood conditions at the site should

be made known to the applicant (or its architects-engineers); and they should be instructed to redesign the facility in accordance with accepted engineering practices that would provide adequate protection against flood hazards or instructed to locate the project to another site.

To accomplish complete coordination and cooperation among the offices in HUD and other agencies in the federal government, Secretary Weaver's order included:

> In areas where flood studies are not available locally, appropriate data should be obtained from other offices in the Department of Housing and Urban Development, the District Offices of the Corps of Engineers and the U.S. Geological Survey. Where necessary, other pertinent data may be acquired through the Bureau of Reclamation and various state agencies.

The purpose of the planning assistance programs is to foster comprehensive planning at all levels—community, metropolitan area, region, and state. They also provide for capital improvements programming and project planning. The grants-in-aid and some of the mortgage insurance programs of HUD specially require that the area in which a project is located must be covered by a comprehensive plan or planning at one or more levels and the project must be consistent with such planning.

The following sections review the financial assistance available for implementing local programs and projects in which flood plains and flood protection measures may be involved.

## PLANNING PROGRAMS

There are two principal planning programs. *Urban planning assistance,* generally known as the 701 program, is in the program coordination and services office of HUD. The Iowa Development Commission and the Office of State Planning and Programming are delegated to administer the program for the state of Iowa. The planning division of the Iowa Development Commission coordinates the actual planning effort of local and regional planning agencies to meet federal requirements.

Specific uses of the program are the preparation by local and state planning agencies of comprehensive development plans, including provision of public facilities, transportation, and long-range fiscal plans. Also included are programming and scheduling of capital improvements. The identification of flood plain lands, their land-use designations and preparation of zoning, subdivision regulations, and other implementing controls can be undertaken as elements in the comprehensive planning process.

Federal grants under the 701 program are available for as much as two-thirds or, in some instances, three-fourths the cost of the research, planning, and reporting work. Who may apply?

1. Cities and other municipalities with 50,000 or under population, counties, and Indian reservations through their state planning agencies.
2. Official state, metropolitan, and regional planning agencies and metropolitan organizations of public officials.
3. Cities and counties in redevelopment areas without regard to size.
4. Official governmental planning agencies for federally impacted areas.

*Advances for public works planning* is a program in the metropolitan development office of HUD. It provides for interest-free advances to assist planning for individual local public works and for area-wide and long-range projects which will help communities deal with their total needs. Planning for all types of public works except public housing is eligible. The applications of this program to flood plain or flood-prone lands are the preparation of plans for local and area-wide drainage systems and other flood protection measures such as levees, floodwalls, bank erosion controls, and the like.

States, municipalities, and other public agencies may apply. An applicant, however, must show intent to begin construction within a reasonable period of time and that financing of such construction is feasible. The public work being planned must conform to an appropriate state, local, or regional plan and be approved by a competent state, local, or regional authority. The advance is repayable to HUD promptly upon start of construction of the planned public work.

## FUNDING FOR PROGRAM IMPLEMENTATION

The programs of HUD for financial aid to implement local programs and project development extend into various areas of regional and metropolitan development.

*Grants for advance acquisition of land,* a program in the metropolitan development office, is for the purpose of encouraging communities to acquire land in a planned and orderly fashion for future

## FEDERAL PROGRAMS FOR URBAN PLANNING AND ASSISTANCE

construction of public works and facilities. An example is the acquisition of lands needed for storm water reservoirs and channel improvements as elements in a regional drainage system.

Local public bodies and agencies may apply. The facility for which the land is used must be started within a reasonable time, not exceeding five years after the grant is approved. Construction of the facility must contribute to the comprehensive planned development of the area. Grants may not exceed the interest charges on a loan incurred to finance the acquisition of land for a period of not more than five years.

*Grants for basic sewer and water facilities,* a second program in the metropolitan development office, is designed to assist and encourage communities of the nation to construct adequate, basic water systems and sanitary and storm sewer facilities to promote efficient and orderly urban growth. A significant special aspect of this program is its contribution to water pollution control. Sewer system installations must comply with standards acceptable to the Federal Water Pollution Control Administration, Department of the Interior.

Federal grants generally pay up to 50 percent of the development cost of basic water and sewer facilities, including the cost of the land. Also included are grants for relocation of individuals, families, business concerns, and nonprofit organizations displaced by activities assisted under the program.

Local public bodies and agencies may apply. The project, however, must be determined necessary to provide adequate water or sewer facilities for the people to be served. It must be designed so that an adequate capacity will be available to serve the reasonable foreseeable growth needs of the area. It must be consistent with a program for unified or officially coordinated area-wide water and sewer facilities systems as part of the comprehensive planned development of the area.

*Open-space land,* also a program in the metropolitan development office, is designed to assist communities in acquiring and developing land for open-space uses. Specific uses of these lands are to provide parks and other recreation, conservation, and scenic areas. Flood plain lands can be acquired, provided the use of the land is not intended to perpetuate the area solely as a flood plain. There must be multiple-purpose use of the land, in that it is accessible to the public and appropriately maintained during significant periods of the year for recreation, conservation, or scenic uses.

Federal assistance has been increased from 20 to 30 percent to a single level of 50 percent of the cost of the land to help public agencies acquire and preserve lands for recreation uses. Twelve percent of the acquisition cost is available for development of the land.

State and local public bodies may apply. Assisted open-space activities must be part of an area-wide open-space acquisition and development program, which in turn is consistent with area-wide comprehensive planning. Developed lands in built-up areas are eligible only if open-space needs cannot be met with existing undeveloped lands.

*Urban renewal* programs of the urban renewal assistance office of HUD assist cities and counties to undertake local programs for the elimination and prevention of slums and blight. Urban renewal is a long-range effort to achieve better urban areas through planned redevelopment of deteriorated and deteriorating areas. This is accomplished by means of partnership among local government, private enterprise, citizens, and the federal government.

Activities and projects in the federal urban renewal program are financed with federal advances and loans, federal grants, and local contributions. Federal grants generally pay up to two-thirds of net project costs—in certain instances as much as three-fourths. The local government's contribution in the amount of at least the required local share of one-third or one-fourth of net project costs may be in cash or noncash grants-in-aid. Noncash grants-in-aid can be through credits for locally financed and developed public improvements or facilities meeting type and time limit criteria of renewal assistance office regulations.

## INCLUSION OF FLOOD PROTECTION MEASURES

Flood protection work, to be eligible as a project improvement expenditure or a local noncash grant-in-aid credit, must be clearly necessary to achieve sound renewal of the project area. The cost must be reasonable in terms of the community objectives gained. Project funds may not be used where flood protection work is the primary objective.

Moreover, the local public agency must submit evidence that the contemplated flood protection work will not:

1. Duplicate work scheduled by other agencies for accomplishment within a reasonable time.
2. Be in conflict with other flood protection work that may be con-

templated by federal, state, or local agencies.
3. Cause adverse effects on other areas without provision of appropriate remedial measures.

Flood protection work as an eligible improvement to be financed with project funds must be secondary to the elimination of slum and blight. The work must be wholly within a project area and be no more than necessary to protect the proposed land uses in that area. Flood protection work for a flood plain area that extends beyond the project boundaries, however, is eligible for noncash grant-in-aid credit to the extent it benefits the project. The percent benefit is based on the area of the flood plain in the project divided by the total area protected by the improvement.

The proposed reuses of cleared land in the project area requiring protection must have special significance to the community. The proposed reuses may call for redevelopment because of location and lack of alternative sites with similar factors or because treatment to halt deterioration in a rehabilitation and conservation section within the project area is dependent on the protection proposed.

A storm sewer system improvement within the project area and such other flood protection measures designed as complete and separate improvements may be eligible. Such flood protection measures may be levees, flood plain landfills, floodwalls, revetments, or bank erosion control. Retaining walls and bulkheading are also eligible when they are essential parts of the flood protection.

Typical renewal projects in which flood protection work has been installed or is contemplated in Iowa and elsewhere are: River Hills Project, Des Moines, Iowa; Westfield-Virden Project, Waterloo, Iowa; Central River Front Project, Cincinnati, Ohio; and Upper Levee and River View Projects, St. Paul, Minnesota.

## OTHER ASSISTANCE PROGRAMS

*Federal Housing Administration* (FHA) programs cover insurance of mortgage loans made by private lenders to finance the purchase of homes, the construction of rental housing, and rehabilitation of dwellings. The FHA also has a program of mortgage insurance for land development. This covers the purchase of raw land, the installation of land improvements for new subdivisions, and complete new communities.

The objective of the land development program is to assist private enterprise in developing land for residential and related uses in a manner that is orderly, economical, and consistent with sound area-wide and community growth. There are two significant special requirements of the program:

1. The area in which a land development is located must be covered by comprehensive planning for the area.
2. The planned subdivision or new community must be consistent with sound land-use patterns and comprehensive planning for the area.

The maximum mortgage amount outstanding for any one land development undertaking at one time is $25 million. The maximum repayment period is seven years, or may be longer for privately owned water or sewerage systems. The ratio of loan to property value is 75 percent of the FHA estimated value of the developed land, or 50 percent of the estimated land value before development and 90 percent of estimated development costs, whichever is less.

**Planned-Unit Developments**

The FHA program for *planned-unit developments* with homes associations is particularly suited for land developments in which the blue-green technique of employing stormwater retention ponds for multiple-purpose drainage and recreation is used. It encourages cluster planning for detached homes, townhouses on the green, and other planning innovations. FHA standards and guidelines give great flexibility to the design profession and building industry in planning of unified residential developments with common open areas.

FHA planned-unit developments apply to any residential development which (1) is a subdivision of land into lots for use predominantly for owner-occupied homes with home mortgage financing such as FHA 203-b and (2) has common property comprising an essential or major element of the development. Ownership and maintenance of the common property is by an automatic membership homes association. This is an incorporated, nonprofit organization operating under recorded land agreements. The common property and facilities in a development also can include those that are publicly owned and operated.

With such a development the benefits to be derived from the

common properties will be reflected in the FHA appraisals of the homes and thus in long-term low-interest home mortgages. This makes it economically possible to create such a development complete with improved common properties and to offer the home properties favorably in a competitive market.

Guidelines governing site-planning standards for FHA multifamily dwellings apply to planned-unit developments under FHA's home mortgage program. Gross acreage instead of net acreage is now used as the land-use measurement. It includes all the open space which benefits the poject, both on-site and off-site, thus more accurately reflecting land-use intensity and project livability. The traditional FHA "must" for vehicular access to every home property does not apply in a planned-unit development if it is accepted in accordance with FHA land planning bulletin No. 6, "Planned-Unit Development With a Homes Association." Detailed plot-planning standards are also set aside.

The common areas and facilities in a planned-unit development can be integral parts of or augment community facilities requirements for the area covered by a general development plan. Of particular significance is enabling control and management of open spaces that encompass flood plain lands or lands subject to stormwater ponding in a residential development. Many localities now have zoning, subdivision regulations, and other applicable ordinances governing establishment and regulations for planned developments. Some have incorporated in their ordinances FHA's planned-unit development or similar standards and guidelines.

## SUMMARY

Field operations of the programs of HUD are administered in seven regional offices. The midwestern states are in Region IV—Iowa, Illinois, Indiana, Michigan, Minnesota, Nebraska, North Dakota, Ohio, South Dakota, and Wisconsin. Region IV offices of HUD are located at Chicago, Illinois. A regional administrator supervises and manages the field operations. There are 14 FHA insuring offices in Region IV. The office serving Iowa is located in Des Moines, headed by a director. Questions concerning the programs of HUD may be directed to the regional office; if the inquiry concerns a program of FHA, the FHA insuring office may be consulted directly.

The programs of HUD will be of considerable value in flood

plain planning and in achieving a sound flood plain management program. Coordination of land uses for the flood plain and flood-free lands in all areas will be required. This will benefit the community, the region, and the nation.

**REFERENCES**

1. Office of the President of the United States. *Evaluation of flood hazard in locating federally owned or financed buildings, roads, and other facilities and in disposing of federal lands and properties.* Exec. Order 11296. 1966.
2. U.S. Department of Housing and Urban Development. *Programs of HUD.* HUD IP-36. Washington: USGPO. 1967.

# 11

# THE INTERACTION OF URBAN REDEVELOPMENT AND FLOOD PLAIN MANAGEMENT IN WATERLOO, IOWA

John R. Sheaffer

MANY of the nation's flood plains contain extensive areas which have been urbanized.[1] Within these problem areas it has been common for agencies to pursue programs that mitigate the effects of flooding without considering many of the interrelated land-use and development problems. Recent policy changes require that urban goals be considered in flood plain management programs. Therefore it is desirable to relate flood plain management efforts to those urban efforts concerned with renewal and redevelopment.

## BACKGROUND INFORMATION

The interrelation of flood problems and other urban deficiencies can be illustrated by an analysis of Waterloo, Iowa; a city of 75,000 that is bisected by the Cedar River (locally known as the Red Cedar

JOHN R. SHEAFFER is Research Associate, Center for Urban Studies, University of Chicago, Chicago, Illinois.

River). The flood hazard area is illustrated in Figure 11.1. Approximately $91 million of urban improvements have been located on the flood plain and average annual flood damages have been estimated at $1,221,651.[2] In addition to the flood problem, however, and approximately coterminous with it, was a blight belt that extended diagonally through the city on both sides of the Cedar River.[3] Unplanned and uneven growth in this area presented an impediment to the smooth flow of traffic in the growth pattern—the streets were narrow and there was a lack of adequate parking. New arteries were needed to permit traffic to circulate, and spaces for parking were required.

The severe land-use mix of factories, homes, schools, and commercial facilities presented safety and health factors. For example, there were unguarded railroad crossings; school children crossing railroad tracks and heavy traffic arteries; poor driver and pedestrian visibility; and the introduction of noises, odors, gases, and smoke from heavy industry into a mixed residential section. An air pollution problem of severe magnitude was developing. From the standpoint of urban facilities there were (1) inadequate storm and sanitary sewerage systems; (2) streets which in general were below acceptable standards, many lacking paving, lighting, curbs, and gutters; and (3) sidewalks which were discontinuous and in a state of disrepair. Because of the existing substandard development and lack of facilities, similar and dependent establishments were unable to generate their own beneficial influence.

The river floodway and water quality also had deteriorated. The introduction of industrial and municipal wastes and heat from power generating had deteriorated the water quality. The banks were marred with dumping and unsightly fill. More significant, perhaps, were the sluggish backwater areas along the main channel, which contained prolific algae blooms and served as mosquito-breeding areas. Because of these conditions, recreational use and enjoyment of the river have been waning at a time characterized nationally by spiraling increases in demand for river access areas.

Waterloo had a flood problem, but it was contained within an urban setting. Associated with it were acute land and water resource management problems which, although not as dramatic as flooding, were potentially of greater economic importance to the area.

## LOCAL FLOOD PROTECTION

Relatively frequent flood threats have resulted in the formulation of a plan of improvement for local flood protection.[4] The elements of

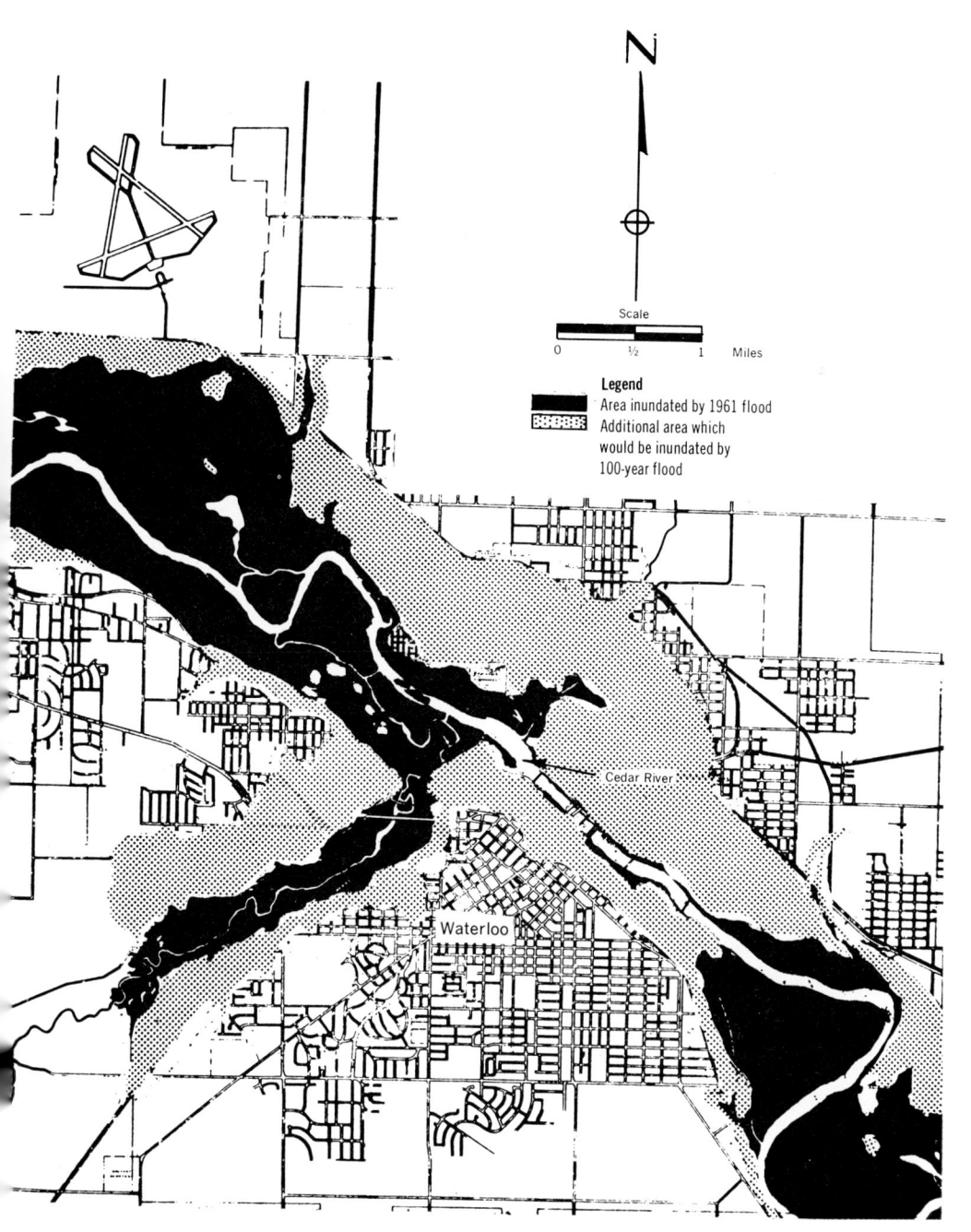

Fig. 11.1. The flood hazard in Waterloo, Iowa, showing the area inundated by the 1961 flood of record and the 100-year flood.

the plan, which can be categorized as a traditional local flood protection effort, are presented in Figure 11.2. The major features are levees and floodwalls along the banks of the Cedar River and Black Hawk Creek to protect against the 100-year flood. Drainage behind these barriers was to be accommodated by a series of pondage areas and pumping stations. In essence, it was a single-purpose public construction solution to control the floodwater, but it would do nothing to mitigate the myriad other problems also present.[5] The net effect of the program could be to convert a blighted urban area subject to flooding to a blighted urban area subject to less frequent flooding.[6]

## COMMUNITY REACTION TO A FLOOD CONTROL PROJECT

A record flood on March 28, 1961, ignited community interest in "doing something" about the flood problem and a flood control committee was activated. As the community leaders viewed the situation in its broad perspective, however, there was a growing awareness of the accompanying urban and water renewal problems. Therefore, on March 21, 1962, a Survey and Planning Application (Iowa R-7C) was submitted to the regional office of the Urban Renewal Administration. This application was submitted under the Housing Act of 1949 as amended and proposed a joint attack on both the urban and water problems (including flooding). Because the application proposed a broader program than had heretofore been considered in urban renewal efforts, a lengthy discussion between the city of Waterloo and the URA took place. On October 8, 1962, an attachment to the initial application was submitted, which stated:

> The City of Waterloo has considered the Westfield-Virden Area as a valley of blight adversely influenced by factors which are beyond its control to remedy. While public funds can be justifiably spent and have been invested in other sections of Waterloo, this project area has been victimized, by its location and by the firmly rooted factors of blight, from receiving the general growth impetus available elsewhere throughout the City. A concerted and effective program of Urban Renewal could salvage these neighborhoods. The flood of 1961 only pointed up factors that had long been in existence and the haunting influence and the firm possibility and near certain likelihood of additional and more frequent disasters remains as a spectre of blight.

Following a public presentation of the Corps of Engineers local flood protection proposal, the Waterloo Flood Control Committee at a special meeting on November 13, 1962, reacted by stating, "every

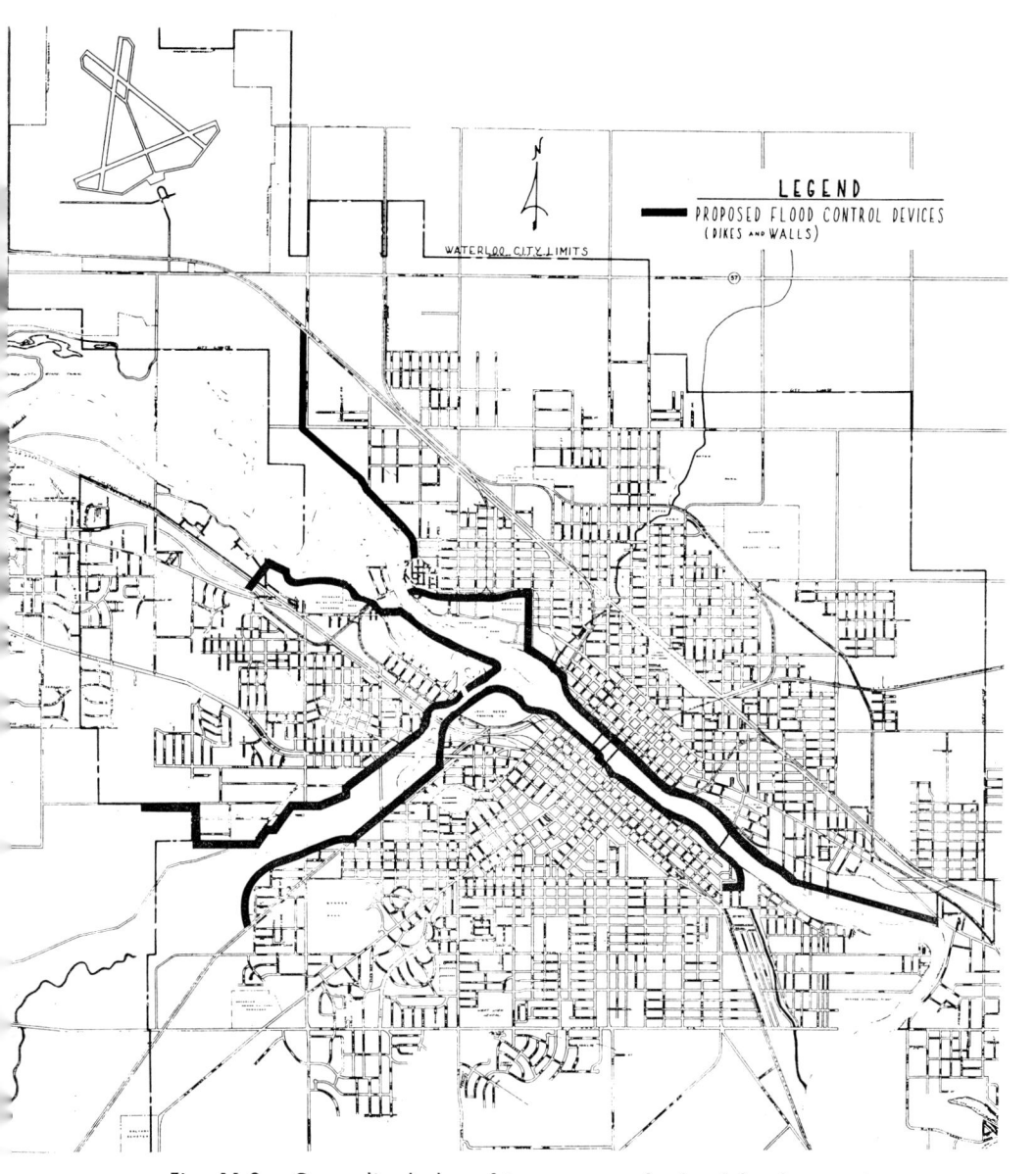

Fig. 11.2. Generalized plan of improvement for local flood protection, Waterloo, Iowa, which was recommended by the Rock Island District, U.S. Army Corps of Engineers.

effort should be made to bring in a flood control and redevelopment program through Urban Renewal." It was felt that the Corps program "was too severe and certain long-term detrimental aspects of it outweigh the benefits . . . the plan was not locally acceptable. The physical aspects of the Corps' plan produced a negative reaction." The main concern of Waterloo was the preservation and reclamation of the esthetic and environmental quality of the Cedar River valley, a typical goal which is becoming more prevalent nationally.[7] Nevertheless, there was a belief that flood control was absolutely essential and that the Corps of Engineers should be an active partner in the urban redevelopment effort. A decision was made to join two major but separate national domestic efforts—flood control and urban redevelopment—into an integrated urban program. The local renewal agency changed its name to the Department of Flood Control and Redevelopment. The community rejected the public-construction single-purpose management strategy and sought a complex metropolitan-area water-management strategy in its place.[8]

## URBAN RENEWAL OPPORTUNITIES

The major part of the flood plain in Waterloo had long been committed to unplanned urban use. Urban renewal offered an opportunity to achieve a proper use of the flood plain. With respect to traffic, a thoroughfare system and circulation system could be constructed with adequate lighting and pedestrian facilities. Also, an adequate storm and sanitary sewerage system could be installed.

Through clearance, blight and mixed uses could be removed and the remaining sound development could produce a beneficial influence on the surrounding area. New sites could be made available for development. Finally, through a comprehensive program of land and water renewal, recreational opportunities could be enhanced. Water-oriented recreational facilities could be expanded, and the scenic riverfront park and recreational areas extended.

Through the application of urban renewal in its broadest and profoundest sense—which encompasses flood plain management—the city of Waterloo could eradicate slums and blight, create new public and private uses, and restore to productive use the diagonal belt of deterioration which bisected the city. A major part of the city could be recovered, capping a century of concern.

The attainment of these goals would permit the dramatic meeting of land and water to be a focal point again; the city once more would face the river and its residents could enjoy the scenic and esthetic vistas of the valley and the open space provided by the water areas. Through such planning even the water areas would take on increased functional importance as water supply, recreation, and floodwater conveyance features.

## A METROPOLITAN-AREA, WATER-MANAGEMENT STRATEGY

The desire of the city of Waterloo to implement the comprehensive approach—land and water management in an urban setting—through a joint flood control and urban redevelopment program presented some problems. The formulation of a plan for urban redevelopment concluded that to promote a more effective and desirable quality of land use and to enhance the marketability potential for land that would have to be redeveloped, the Corps plan for flood control had to be modified. The modifications within the renewal area are illustrated in the site plan shown in Figure 11.3 and are discussed briefly in the following paragraphs.

Along the west bank the mixed deteriorated area is to be cleared and the site raised by "rolling back" the riverbank, eliminating some island obstructions, and using the material excavated to elevate the bank above the estimated 100-year flood level. This area, under the plan for local flood protection, would have been shut off from the river by a levee and would still have been subject to inundation from the catastrophic flood. Through the renewal program it takes on the air of a prime industrial area serviced with adequate traffic arteries, including a scenic riverfront drive and vista of the river. A side effect of this modification would be a greater water surface area for recreation and an increased channel cross section. As a part of this modification the reach of Black Hawk Creek that flows through the industrial property of the John Deere Company is to be relocated to enter the Cedar River upstream from its present entrance. This relocation proposal triggered interest in industrial development in Waterloo by providing a well-located, flood-free site for the John Deere firm's expansion. The reservation of some area for filling with industrial solid wastes incidentally would mitigate the solid-waste disposal problem faced by the local industry.

On the east bank, Exchange Park and Cedar River Park are at present split by a strip of generally deteriorated residential development one block wide. The flood protection plan called for a levee up to 17 feet in height to border the river and to protect the area. If

**WESTFIELD-VIRDEN urban renewal project**

through renewal this deteriorated area could be cleared and the sound houses relocated on higher sites, the levee would not be necessary. In its place a low embankment approximately 3 feet in height along the railroad right-of-way and tying into high ground at the "Knolls" and Fairview Cemetery would suffice. This modification would unite the park areas and preserve their scenic front door as an alternative to a 17-foot-high barrier that would have shut off the river from the park.

Another modification concerns the ready-mix concrete plant located upstream from the Mullan Avenue bridge. Here the riverfront has been defaced by washings from the trucks, unsightly fill, and stockpiles of material. Because of lack of space between the existing development and the channel, the flood protection project called for a concrete floodwall along this reach of the river. From a planning viewpoint, a ready-mix plant generally is incompatible with central business district activities. Because of this conflict the plant would be acquired and the activity relocated to a more suitable site. As a result the costly reach of floodwall would no longer be necessary. This modification also would provide the necessary right-of-way for the proposed riverfront drive (a needed traffic artery) and would permit the continuation of the beautification of the riverbanks with green areas.

One other modification should be mentioned. As a result of the comprehensive approach, the new recreation center constructed between 1st Street and 2nd Street on the riverbank was designed as a floodproofed building which also serves as a reach of floodwall. This building preserves the waterfront while assisting in providing flood protection.

As a result of these modifications, which enhance significantly the urban development aspects of the flood protection project, the estimated costs are reduced by more than $.5 million.[9]

The permanent water level through the project area currently is maintained by a low head dam at Park Avenue. The fixed low crest of this dam does not provide sufficient water depth in the upstream reaches for recreational boating in the summer, and in addition it acts as a sediment trap. Because of its effect on the project, this dam would be replaced with one containing adjustable gates that could be set at a higher elevation to provide an enlarged permanent pool and also could be withdrawn periodically to minimize sediment deposition. These deposits of sediment not only restrict water recreation activities on the river but also tend to impede the flow of water. The design of the proposed new dam would permit the river to flush out the sediment during flood periods.

The modifications formulated in the redevelopment plan were not conceived as opposition to the Corps project, but as an expansion of a single-purpose public-construction project into a metropolitan-area multipurpose and multimeans effort involving both public and private interests. In so doing, flood control and urban renewal efforts work together in the realization of urban goals. Favorable reactions to the proposed modifications exceeded all expectations. The local decision makers sensed the potential benefits and began to organize to promote the proposals. The riverfront drive, long considered to be politically unfeasible, suddenly became a reasonable consideration. Through the aggressive leadership of the local urban renewal administrator, efforts to implement the expanded flood control–urban renewal program were launched.

As an indication of the magnitude of activity, a list of the events in the formulation of the metropolitan water resource management strategy at Waterloo, Iowa, follows:

| Date | Event |
| --- | --- |
| March 28, 1961 | Record flood on the Cedar River. |
| March 31, 1961 | Waterloo declared a disaster area. |
| March 21, 1962 | City submitted to Housing and Urban Development (HUD) a Survey and Planning Application, Iowa R-7C. |
| October 4, 1962 | Generalized description of a proposed flood protection project provided by the Corps of Engineers. |
| October 8, 1962 | Attachment to Survey and Planning Application submitted. Preliminary meeting between Corps and city held. |
| November 13, 1962 | Recommendations of Waterloo Flood Control Committee. |
| November 28, 1962 | Meeting between Corps and city. |
| December 4, 1962 | City resolution—Corps plan found generally acceptable but city desired to become an active partner. |
| September 4, 1963 | Meeting held with local interests to develop recreational potential. |
| September 11, 1963 | Meeting held to develop the flood control and water resource aspects of the project area. |
| March 12, 1964 | Application submitted to the National Rivers and Harbors Congress. |
| May 5, 1964 | Discussion between Corps and urban renewal consultants to resolve differences. |
| May 26, 1964 | Submission of Part 1 of the application for the Westfield-Virden urban renewal project. |

| Date | Event |
|---|---|
| September 21, 1964 | Bridge removal discussed with HUD. |
| October 15, 1964 | Public hearing on the Corps project. |
| April 15, 1965 | Submission of Part 2 of the application and execution of the loan and grant contract between the city of Waterloo and HUD. Project was initiated. |
| December 31, 1967 | Acquisition of all parcels in the Westfield-Virden urban renewal project area. |

The reaction of the district engineer of the Rock Island District, Corps of Engineers, was that the two efforts—flood control and urban development—tended to enhance each other when blended.[10] The text of his letter read:

> In the meeting yesterday, we discussed the relation between the flood control project for Waterloo and the Westfield-Virden Urban Renewal project.
> From the review accomplished, it is apparent that the two projects can be blended so the flood control objectives can be achieved and the urban renewal objectives attained, with each project tending to enhance the other.
> The meeting with you and other representatives of Waterloo yesterday was a productive and enjoyable one.

## IDENTIFICATION OF BENEFITS

The modifications proposed in the management effort have the potential for producing benefits in addition to those which would have been realized by the single-purpose flood protection levee project. These benefits are summarized below.

The program would make available approximately 150 acres of prime industrial land with a panoramic view of the river, elevated above the estimated 100-year flood stage. Incidentally, a large part of this land has been purchased by the major industry in the city as sites for new plants. Residual flood damages within the filled area also would be reduced, since there is less potential for damage from the occurrence of a catastrophic flood; if the design elevation is exceeded, the area would be affected by low flood stages only.

The recreational potential of the river would be enhanced by improved water quality and expanded surface area. The dredging of material from the existing channel and flood plain and the removal of some islands would add 100 acres of recreation water.

A mosquito-abatement benefit would be realized, for the elimination of backwater sloughs and some ponding areas would destroy their breeding areas. The provision of a site for industrial waste disposal adjacent to the industry providing large volumes of wastes greatly reduces the cost of solid waste disposal.

The urban esthetics would be preserved by designing the flood control project so that there would be minimum disruption of the natural environment.[11] Where the environment has been marred, it will be reclaimed through urban redevelopment.

These benefits have been quantified in the urban renewal survey and planning report and, on an annual basis, total $212,000. Thus the expanded strategy increased the flood control project benefits by 27 percent while at the same time reducing project costs. More significant, however, are the long-term economic effects of the program. As reported in the Waterloo newspapers, signs of rebirth are in evidence in the center of Waterloo. A record amount of plant expansion, $26 million, was announced. The major employer in the city, the John Deere Company, has purchased 125 acres and will construct a $20 million electric foundry in the filled industrial area. This decision goes a long way toward assuring the economic well-being of the area. It was recognized that if the John Deere firm had decided to build at another city, this would have been a severe economic blow to Waterloo's future.

## CONCLUSIONS

The experience in Waterloo illustrates the desirability of joining two major domestic programs, urban redevelopment and flood control, into a unified effort. The success of the effort is presented in the 1967 urban renewal report, *The Banks of the Cedar Turn to Green . . . and Gold*. It is observed, "to the City of Waterloo comes a new sense of direction, a new spirit of vitality, a new surge of confidence." Actions under way testify to the validity of the observation.

The complex metropolitan-area land and water management program faced many agency and organizational obstacles; but because the potential benefits were perceived, local decision makers tenaciously pursued its implementation. When confronted with an obstacle, the decision sought was on action to overcome it, rather than using the difficulty as an excuse to regress to a single-purpose public-construction effort.

Although a complex and relatively unused management strategy was employed at Waterloo, the scheduling of program efforts calls for a 1971 completion date. Thus it appears that a program can be formulated, discussed, and implemented all within a decade, a considerably shorter time span than the average 20 years for a flood control project.

In Waterloo a healthy combination of "imagineering" and local capabilities succeeded in uniting two major domestic programs. In doing so, the city also succeeded in accelerating the implementation of the combined program.

## REFERENCES

1. In G. F. White, et al., *Changes in urban occupance of flood plains in the United States* (Chicago: Univ. of Chicago Press, 1958). Over 1,200 cities were estimated to have flood problems. A recent listing by the Corps of Engineers of urban areas over 2,500 with flood problems showed that virtually all urban areas had some flooding or drainage problems.
2. Estimated in U.S. Army Corps of Engineers, *Interim review of reports for flood control on the Iowa and Cedar rivers, Iowa and Minnesota* (Rock Island: U.S. Army Eng. Dist., 1964).
3. The discussion of urban blight is based on findings presented in Urban Renewal Administration, *Urban renewal rept.*, pt. 1, Westfield-Virden Project, Iowa R-7C. 1964.
4. U.S. Army Corps of Engineers, *Interim Review*.
5. See G. F. White, *Strategies of American water management* (Ann Arbor: Univ. of Michigan, 1967), in which six types of strategy are presented. Single-purpose construction by public managers is one of the simplest and oldest strategies.
6. Protection levels were for the 100-year flood, and $422,351 or more than one-third of the average annual losses, would continue as residual damage.
7. Environmental quality is becoming a major consideration in water management efforts; see Committee on Water, *Water and choice in the Colorado basin* (Washington: Nat. Acad. of Sci., 1968).
8. See G. F. White, *Strategies*. This complex strategy is discussed in Chapter 6.
9. Urban Renewal Administration, Iowa R-7C. 1964.
10. A letter from Col. Richard L. Hennessey, District Engineer, Rock Island Dist., to Mayor Ed Jochumsen, Waterloo, Iowa, May 6, 1964.
11. Preservation of the natural environment is now recognized as a first level alternative (objective). See U.S. Army Corps of Engineers, *Susquehanna River basin study plan: A review of alternatives* (Baltimore: Corps of Engineers, 1967), p. 5.

# 12

# THE ROLE OF OPEN SPACES IN FLOOD PLAIN MANAGEMENT

Donald K. Gardner

THE ADMINISTRATIVE, supervisory, and planning personnel in the field of parks and recreational opportunities share with the flood plain manager an interest in the streams and valleys of our state. This interest concerns not only federal and state agencies but people from within and without the state and from all over the country. All are working to solve a problem that is most important to elected city officials, who spend substantial amounts of money on flood control and in fighting floods in urban areas.

This chapter will be devoted to the role of open-space acquisition in programs of flood plain management. Of importance in this discussion is the need for a balanced allocation of land use in urban areas, and it will be shown that open-space needs are compatible with concepts of optimum flood plain utilization and efforts to minimize flood plain occupancy by the more intensive uses.

DONALD K. GARDNER is Commissioner of Parks and Public Property, Cedar Rapids, Iowa.

## OPEN-SPACE ACQUISITION AND GROWTH OF URBAN PARKS

As a long-time resident of the city of Cedar Rapids and a city employee or official since 1941, this author has performed most if not all of the jobs which are encountered in a park system. This includes emergency efforts during floods and a great many hours spent at miscellaneous tasks unrelated to parks and recreation. All too frequently this has included filling sandbags to protect low-lying, run-down residential and business areas that perhaps should never have been built in the first place. Following the floods of the early 1960's, the city of Cedar Rapids is taking a new look at the flood problem and making a new attack on it.

The population in Cedar Rapids has grown rather phenomenally in the last fifteen to twenty years. A 27 percent increase was experienced between 1950 and 1960, the 1960 population being 92,035. An official census in 1965 gave a population of 103,545. Linn County, of which Cedar Rapids is the county seat, had a 1960 population of 136,900. This represented a 31 percent increase since 1950 and indicated that suburbs of Cedar Rapids were growing even more rapidly.[8] The population of Cedar Rapids is now estimated to be almost 110,000, a community of considerable size and stature and now the second largest city in the state.

There were 780 acres of parkland in the city in 1955. By 1967 this had been increased to 2,200 acres, almost a threefold increase. It is clear that the city has been deeply involved in the land acquisition business during the last few years.

City officials also discovered early in the acquisition program that land values were spiraling rapidly. This has also been true nationally.[9] There is a limit to how much the city can spend as well as to how much land can be taken out of production, out of commercial use, out of industry, and out of residential use for park purposes.[1]

Too frequently the undeveloped areas are considered unused and unproductive. Yet open space is a needed partner in urban land-use planning and in allocating land to the respective urban uses. Four major elements have been included by Clawson in the program for reservation and preservation of open space within and near urban areas.[1]

1. The kinds of open space desired must be defined as carefully as possible.

## ROLE OF OPEN SPACES

2. Reasonable space standards under differing circumstances must be formulated for each.
3. Imaginative yet realistic programs for using each type of open space must be developed and carried out.
4. In the process and as a direct objective, public-interest groups must be developed and encouraged to support the continuance of the open space during the political struggle that will arise from time to time.

A system of classifying land and water areas which associates the recreation uses to the most suitable areas has been suggested.[3,11] Six categories were named:

1. High-density recreation areas for intensive use.
2. General outdoor recreation areas.
3. Natural environment areas.
4. Outstanding natural areas.
5. Primitive areas.
6. Historic and cultural sites.

Because of the tremendous cost of land acquisition, special opportunities should be used to an advantage. Flood plain lands fall within this special concept.

Cedar Rapids has acquired most of the major regional areas that will be needed for park and recreational purposes. The city will continue to acquire recreational sites, especially neighborhood sites for active uses primarily associated with the schools. Our big move, however, is obtaining the areas to be needed as open spaces for the increased leisure-time activities supposedly lying ahead. Leisure time is fast becoming one of the nation's pressing social problems, as idle hands and idle minds have a tendency to create work for our social agencies. These new areas most likely will be in the flood plains. This is land the city can best afford to buy. It is land that can be removed from the tax rolls, for a great deal of it is neither being used nor heavily taxed. It is land that people are going to need for open-space uses.

## GREENBELT POSSIBILITIES ALONG THE STREAMS

Cedar Rapids is blessed or cursed, depending upon the relative level of flooding, with three major waterways coursing through the city. The Cedar River, known locally as the Red Cedar, divides the city from the northwestern corner to the southeastern corner. Indian Creek meanders north to south along the east boundary, and Prairie Creek flows west to east along the southern boundary. This provides an excellent opportunity to create a greenbelt—an emerald necklace—for the community.

Except for the commercial downtown area located on both banks of the Cedar River, most of our flood plains are still open and available. However, we are beginning to get some minor commercial and industrial encroachment along Prairie Creek. Much of the flood plain land is forested and is very attractive and scenic. It would be entirely satisfactory for park type activities. Because of the flooding situation, land values have remained low, and taxes derived from these lands are rather negligible.

In this program it is proposed that a greenbelt be acquired that encompasses at least the design floodway, natural or improved, and includes in many areas a major portion of the flood plain. Hiking and horseback trails would be developed in these areas. If interest warranted, bicycle trails could also be included. At Cedar Rapids it is possible to develop a trail which would extend from the north edge of the city along Indian Creek, south to the Cedar River, thence northwesterly along the Cedar River to the mouth of Prairie Creek, and finally in a westerly direction along Prairie Creek to our west city limits. This particular trail and greenbelt would extend 25 miles and would encompass from 2,000 to 3,000 acres of land. A city map in Figure 12.1 shows the location of the parkland and proposed greenbelt in Cedar Rapids.

Where high population densities were anticipated, this greenbelt would be expanded to include areas for playgrounds, athletic uses, and major picnic centers. The bayous and backwaters that are created in these flood areas make excellent ice-skating ponds, and some are already used for this purpose. Several picnic shelters have been constructed in the flood plain and were so designed that even though inundated once or twice a year no harm is done. A simple cleanup job puts the buildings back into operation. Figures 12.2 and 12.3 illustrate the open-use type of shelter under normal and flood conditions.

A golf course was constructed in 1959 in the flood plain of Prairie Creek. In the nine years it has been in operation, only three weeks of play have been lost due to flooding. Although the greens-

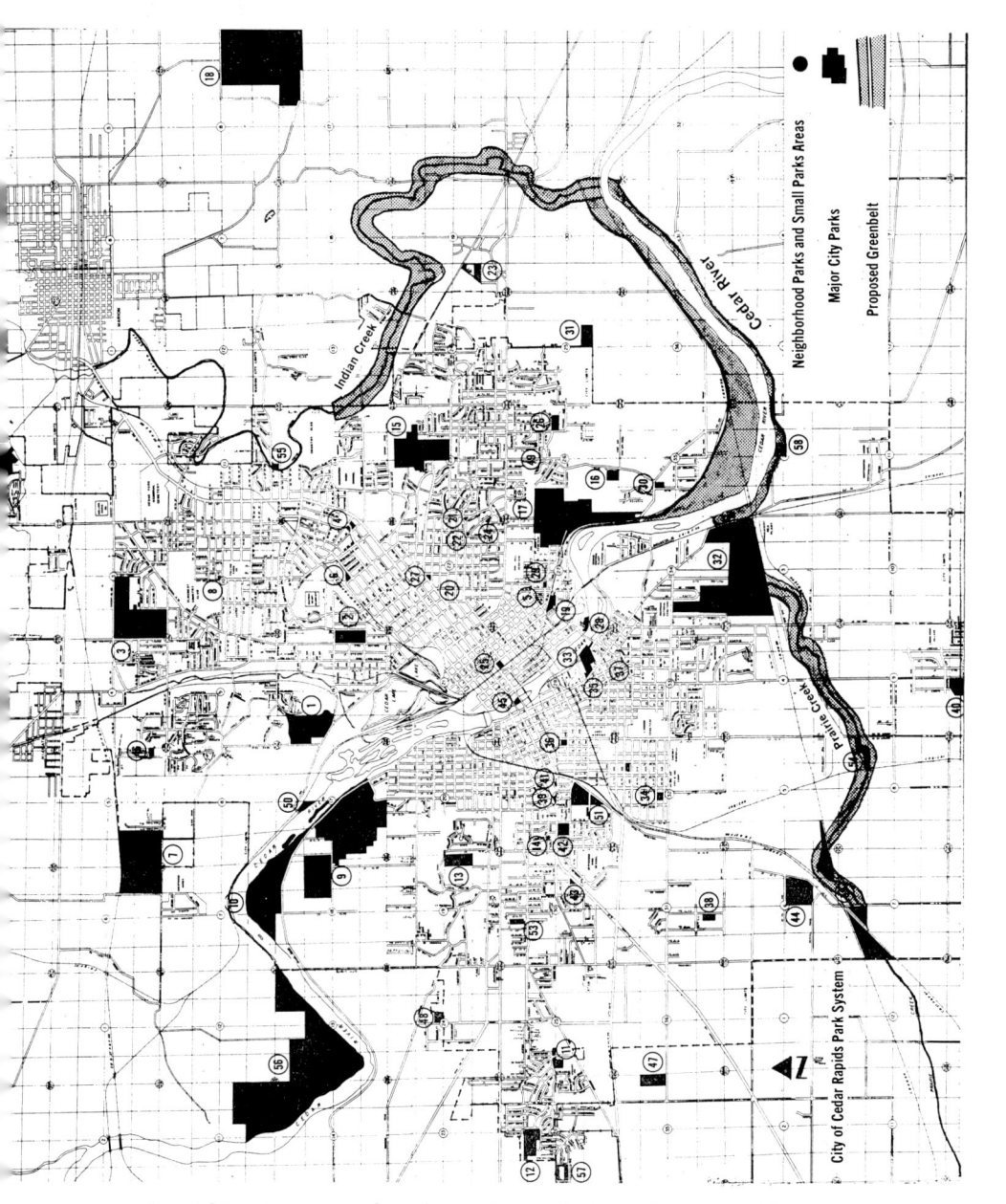

Fig. 12.1. Location of parks and greenbelts in Cedar Rapids, Iowa.

Fig. 12.2. A park shelter designed as an open structure to minimize effects upon flood stages and flood damages.

keeper is quite perturbed about the danger to the greens, no permanent damage has been suffered. This appears to be a very compatible use of the flood plain.

The open-space development the city could make of this land would result in a great amount of public use. Flood fighting measures are minimal, and the cost of cleanup is minor. In existing flood plain areas where homes and businesses have been permitted, there is a constant problem of temporary protection, flood fighting, and emergency services for the displaced families. It far better to leave these areas in open-space use than to pay the price in suffering and dollars that is necessary when the flood plain is occupied by incompatible uses. The use of the river and flood plain for recreational uses is shown in Figure 12.4.

## LOCAL COORDINATION PROBLEMS

Several residents in Cedar Rapids have approached the city council with the problem of having acquired residences in the flood

**ROLE OF OPEN SPACES**

Fig. 12.3. A park shelter during a flood on the Cedar River.

plain unknowingly; that is, there was no way for them to know that the quiet stream near their home would one day surround their house and force them to move. Some way must be provided to record this information on abstracts, so that new buyers are made aware of the flooding potential. This is of special importance for real-property abstracts of all flood plain lands, both developed and undeveloped.

Cedar Rapids has been fortunate in regard to flood plain technical information. Through the cooperation of the city engineering staff, the Iowa Natural Resources Council, and the U.S. Army Corps of Engineers the flood problems have been studied, flood plain technical information is available, and the encroachment limits have been definitely established. Studies were made first of Indian Creek and Dry Creek in 1964.[2,6] Prairie Creek studies were completed in 1966.[7] The Cedar River was studied separately, and a special report for Linn County was published in 1967.[5] This gives the city administration a wonderful working tool with which to accomplish the mission of controlling developments along its streams.

Also of tremendous aid in the program has been the cooperation of the Department of Housing and Urban Development through their open-space land-acquisition program.[12] This provides up to 50 percent of federal aid as a cash grant for the acquisition of open space that is in accord with the regional planning and land-use program. This cooperation between all levels of government and all agencies is giving the Cedar Rapids program a tremendous boost, and within the next ten years the city should be able to complete the greenbelt.

There is another federal program that makes monies available for this type of land acquisition and, although the city of Cedar Rapids has not utilized it, the Linn County Conservation Board has had a program with that agency. The Land and Water Conservation Fund program of the Department of the Interior Bureau of Outdoor Recreation makes federal funds available to the states and gives the states the prerogative of allocating these funds to the political subdivisions.[3,10] In the state of Iowa approximately 50 percent of the funds are retained by the state, and the other half are made available to the political subdivisions. There are planning requirements in this program to ensure that the proposed park and open-space programs are meaningful.[4,11] Officials of Cedar Rapids anticipate that the county conservation board will pick up the flood plain acquisition program where the city leaves off and extend this program throughout the county.

## SUMMARY

Predictions have been made as to what data processing and automation would do to the normal workweek. Some forecast that from 2 to 10 percent of the people will be doing the work while the rest of the population will not be gainfully employed. Others point out that since 1940 a wage earner's leisure time has increased nearly 20 days annually. If these predictions and estimates are even approximately accurate, a great need can be anticipated for public recreation areas for the use of this vast amount of leisure time. It is believed the logical solution to these land needs lies within the flood plains.

At a recent Chamber of Commerce dinner in Cedar Rapids, it was impressive to observe the passing of the gavel when the new president turned to the outgoing one and said, "Hats off to the outgoing for a job well done, and coats off to all for the job that lies ahead." Public support must be gained for programs of land acquisition and development, and community participation in developing sound programs is needed.

# ROLE OF OPEN SPACES

Fig. 12.4. Recreational uses on the Cedar River at Cedar Rapids, Iowa.

As a park-oriented councilman, the author looks upon the flood plain as a greenbelt—an emerald necklace—the lungs that breathe life into the community. It is considered to be Mother Nature's bed, and it is suggested that every precaution be taken to include planning for open-space acquisitions. Industrial, commercial, and residential interests should occupy other areas and should permit the flood plains to be placed in the greenbelt concept.

## REFERENCES

1. Clawson, Marion. A positive approach to open-space preservation. *J. Am. Inst. Planners* 28, no. 2 (1962): 124–29.
2. Iowa Natural Resources Council. *Effects of flood plain encroachment, Indian and Dry Creeks, Linn County, Iowa.* Mimeo. Des Moines. 1966.
3. National Association of Counties. *County action for outdoor recreation.* Rept. of citizens committee for Outdoor Rec. Res. Rev. Comm. Washington. 1964.
4. State Conservation Commission. *Guide for county conservation boards.* Des Moines. 1959.

5. U.S. Army Corps of Engineers. *Flood plain information report, Cedar River, Linn County, Iowa.* Rock Island: U.S. Army Eng. Dist. 1967.
6. ———. *Flood plain information report, Indian and Dry Creeks, Linn County, Iowa.* Rock Island: U.S. Army Eng. Dist. 1964.
7. ———. *Flood plain information report, Prairie Creek, Linn County, Iowa.* Rock Island: U.S. Army Eng. Dist. 1966.
8. U.S. Bureau of the Census. *County and city data book, 1967.* Washington: USGPO. 1967.
9. U.S. Bureau of Outdoor Recreation. *Recreation land price escalation.* Washington: USGPO. 1967.
10. U.S. Congress. *Land and water conservation fund act* (became effective Jan. 1, 1965). PL 88-578. 88th Cong., 2nd sess. 1964.
11. ———. *Outdoor recreation for America.* Rept. of the Outdoor Rec. Res. Rev. Comm. 87th Cong., 2nd sess. 1962.
12. U.S. Department of Housing and Urban Development. *Programs of HUD.* IP-36. Washington: USGPO. 1967.

Missouri River at Blencoe, 1952. Courtesy Corps of Engineers, Omaha District.

**Part 5**

*In the final analysis, the individual counties and cities have to pass the necessary regulations to prevent flood damage, and have to enforce these regulations continuously.*

Francis C. Murphy

# DIRECT CONTROL OF LAND USE THROUGH LAWS AND ORDINANCES

# 13

## COORDINATION OF PLANNING WITH REGULATORY CONTROLS—AN INTRODUCTION

### William M. McLaughlin

THE IMPORTANCE of implementing comprehensive planning programs through effectuation measures is becoming more pronounced, as urban growth accelerates. The expanding urban population is evident even in Iowa where the state's population increase from 1950 to 1960, 5.2 percent, was much less than the national average of 18.5 percent.[4] The population growth of six of the seven counties of greatest population, as shown in Table 13.1, exceeds the state average. Metropolitan areas within these seven counties are Standard Metropolitan Statistical Areas of the U.S. Bureau of the Census.[3]

Linn County experienced the greatest increase, with an impressive gain of 31.3 percent. Cedar Rapids, the county seat, increased in population from 72,296 in 1950 to 92,035 in 1960. The 1965 special census listed a population of 103,545, and a recent estimate by postal authorities was almost 111,000. This represents a 53 percent increase since 1950.

This rapid urban expansion requires more and more land de-

WILLIAM M. MCLAUGHLIN is Planning Director, Planning Division, Iowa Development Commission, Des Moines, Iowa.

151

Table 13.1. Growth of the seven most populous urban counties in Iowa, 1950–1960

| County | County Seat | Population 1960 | Percent increase, 1950–1960 |
|---|---|---|---|
| Black Hawk | Waterloo | 122,482 | 21.9 |
| Dubuque | Dubuque | 80,048 | 12.2 |
| Linn | Cedar Rapids | 136,899 | 31.3 |
| Polk | Des Moines | 266,315 | 17.8 |
| Pottawattamie | Council Bluffs | 83,102 | 19.3 |
| Scott | Davenport | 119,067 | 18.2 |
| Woodbury | Sioux City | 107,849 | 3.8 |

Source: U.S. Bureau of the Census, County and City Data Book, 1967.

velopment. Agricultural and vacant lands are being covered with residential subdivisions, homesites, commercial centers, industrial plants, transportation routes, and municipal utility expansions. Often these lands are subject to serious flooding, and developers frequently do not or will not recognize these problem areas. Planning must keep pace with this growth trend if we are to achieve optimum use of flood plain lands. But of equal or even greater importance are the effectuation measures, including flood plain regulations. Particularly in flood plain development and utilization we need legal tools to implement and enforce the detailed plans for land use and community growth for the public interest. Positive coordination procedures between the planning and regulatory phases will receive major emphasis in this introduction.

## COORDINATION OF PLANNING AND REGULATORY ACTIVITIES IN IOWA

In Iowa, state and federal coordination of regional, county, and city planning activities is administered through the planning division of the Iowa Development Commission.[1] The Urban Planning Assistance Program is provided for and financed under Section 701 of the Federal Housing Act of 1954, as amended.[5] State funds have been allocated by the state legislature to the Iowa Development Commission. Costsharing normally has been two-thirds federal and one-third state and local. Comprehensive plans prepared under the combined federal-state-local program provide the guidance for continued community growth and become the basis or foundation for legal regulations and ordinances under which urban growth can be controlled effectively. However, because flood plain regulation is within the

## PLANNING AND REGULATORY CONTROLS

authority of the Iowa Natural Resources Council, the state water resources agency, additional coordination must be achieved.[2] A positive coordination plan has been established between the planning division of the Iowa Development Commission and the Iowa Natural Resources Council (INRC), through which flood plain development and use can be controlled.

Applications by local and regional agencies for planning assistance under the federal 701 program (now administered by the Department of Housing and Urban Development at the federal level)[5] are made through the Iowa Development Commission. A copy of each application is referred to the INRC for comments and preliminary information concerning existing or potential flood problems. The planning groups are then alerted to specific flood problems and the need for coordinating planning and regulatory measures in flood plain areas. Upon completion of the planning study, a draft copy of the final report is forwarded to the INRC. It is reviewed to ensure conformance to desirable and acceptable measures for control of flood plain developments as specified under state law. Additional coordination between local plan and zoning commissions and the INRC can then be arranged, especially during the implementation phase, to ensure compliance with state flood plain regulatory provisions.

A substantial number of counties, cities, towns and regional areas in Iowa have been involved in the 701 urban-planning assistance program since it began in the late 1950's. A list of the 146 participating cities and towns follows:

| | | | |
|---|---|---|---|
| Ackley | Carlisle | Creston | Fort Dodge |
| Adel | Carroll | Decorah | Fort Madison |
| Afton | Cedar Falls | Denison | Glenwood |
| Albia | Center Point | DeWitt | Greenfield |
| Algona | Central City | Dike | Grinnell |
| Alta | Chariton | Dubuque | Grundy Center |
| Altoona | Charles City | Dunlap | Harlan |
| Ames | Cherokee | Durant | Hartley |
| Anamosa | Clarinda | Dyersville | Hawarden |
| Anita | Clarion | Dysart | Humboldt |
| Ankeny | Clear Lake | Eagle Grove | Ida Grove |
| Atlantic | Clinton | Eldora | Independence |
| Avoca | Clive | Eldridge | Indianola |
| Belle Plaine | Columbus City | Estherville | Iowa Falls |
| Belmond | Columbus Junction | Evansdale | Jefferson |
| Bloomfield | Conrad | Exira | Jewell |
| Bondurant | Coralville | Fairfield | Kalona |
| Boone | Corning | Fayette | Keokuk |
| Burlington | Cresco | Forest City | Knoxville |

| | | | |
|---|---|---|---|
| Lake City | New Hampton | Remsen | Titonka |
| Lamoni | Newton | Rock Rapids | Toledo |
| La Porte City | Northwood | Sac City | Urbandale |
| Le Mars | Norwalk | Sanborn | Van Meter |
| Leon | Oakland | Schleswig | Waukee |
| Lone Tree | Oelwein | Scranton | Waverly |
| Manchester | Onawa | Sheldon | Webster City |
| Manning | Orange City | Shenandoah | Wellman |
| Mapleton | Osage | Sibley | West Branch |
| Maquoketa | Osceola | Sioux Center | W. Burlington |
| Marion | Oskaloosa | Springville | W. Des Moines |
| Marshalltown | Ottumwa | Spencer | West Liberty |
| Mason City | Parkersburg | Stanton | Williamsburg |
| Mediapolis | Pleasantville | Storm Lake | Wilton Junction |
| Missouri Valley | Polk City | Story City | Winfield |
| Mt. Pleasant | Princeton | Sumner | Winterset |
| Muscatine | Red Oak | Tama | |
| Newell | Reinbeck | Tipton | |

The 17 counties (outside of all incorporated areas) and 16 metropolitan and regional areas that have completed planning and zoning studies, or have them in process as of 1968, are listed below. Their locations are shown in Figure 13.1.

| Counties | Metropolitan and Regional |
|---|---|
| Adams | Black Hawk County (Waterloo metropolitan area) |
| Bremer | Central Iowa (communities in and near Polk County) |
| Buena Vista | Calhoun County and communities |
| Butler | Cerro Gordo County and communities |
| Dallas | Cherokee County and communities |
| Floyd | Franklin County and communities |
| Hamilton | Freemont County and communities |
| Hardin | Guthrie County and communities |
| Howard | Johnson County and communities |
| Marion | Jones County and communities |
| Montgomery | Linn County and communities |
| Plymouth | Madison County and communities |
| Poweshiek | Mills County and communities |
| Tama | Northwest Iowa Regional |
| Wapello | Lyon County and communities |
| Washington | O'Brien County and communities |
| Winneshiek | Osceola County and communities |
| | Sioux County and communities |
| | Scott County Regional (Davenport, Bettendorf, and Riverdale with Scott County, Illinois) |
| | Woodbury County and communities |

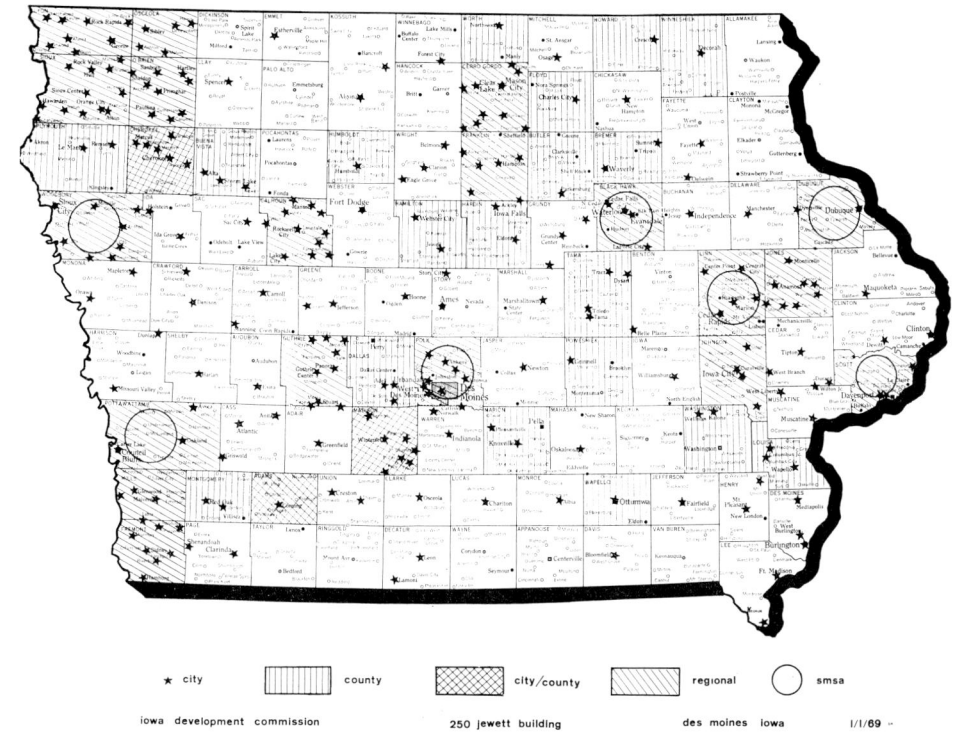

Fig. 13.1. Location map for 701 planning assistance programs in Iowa, 1968.

These numbers indicate, first, a tremendous interest in the planning and regulatory phases of community growth and, second, a great potential for achieving wise planning and use of the flood plains in Iowa. It must be emphasized, in relation to the increased rate of planning, that the corollary regulatory measures are needed as a legal means of reasonably controlling land use so that future development of the community can be enhanced. Then through capital improvement programs, community and regional growth can proceed in a positive manner.

## CONCLUSIONS

In our urban planning program more and more towns, cities, counties, and regions are augmenting and updating their comprehensive plans with statutory regulations to meet changing conditions more adequately. Characteristically, flood plain regulations include such items as zoning ordinances, subdivision regulations, building codes, and health and safety regulations. In our planning work we have found that these statutory regulations do not come about easily. Local officials are often reluctant to adopt legislation which might in some way impinge on the rights of a few individuals, even though the public would benefit. However, federal legislation and the planning and zoning legislation of the Sixty-first General Assembly of Iowa, as well as subsequent legislation, has provided additional authority and assistance in implementing comprehensive plans. The same regulations are also germane to the reduction of damage in areas where flooding may occur.

The various aspects of statutory regulations for flood plains will be discussed in the chapters contained in this section. These efforts should lead to more uniform regulatory provisions which will enhance their adoption and use in local communities and receive more favorable support by the courts.

## REFERENCES

1. Iowa Code. Ch. 28. 1966.
2. Iowa Code. Ch. 455A. 1966.
3. U.S. Bureau of the Budget. *Standard metropolitan statistical areas.* Washington: USGPO. 1967.
4. U.S. Bureau of the Census. *County and city data book,* 1967. Washington: USGPO. 1967.
5. U.S. Department of Housing and Urban Development. *Programs of HUD.* IP-36. Washington: USGPO. 1967.

# 14

# STATE STATUTORY RESPONSIBILITY IN FLOOD PLAIN MANAGEMENT

Clifford E. Peterson

THIS CHAPTER is concerned with state statutory responsibility with regard to flood plain management programs. A brief review will be presented of the history and current application of the principal Iowa statutes controlling some flood plain activity or project which could affect or be affected by a flood event.

## THE NEED FOR LAND-USE CONTROLS

The urgent need for recognition of flood hazards associated with the development of flood plain lands is amply demonstrated by the continued increase in average annual flood losses despite the construction of protective works to control the river in time of flood.[21] These protective works have been constructed at a cost to the general taxpayer of billions of dollars. The expenditures of two federal agencies alone, the U.S. Army Corps of Engineers and the Soil Conservation

CLIFFORD E. PETERSON is currently on leave from the Iowa Natural Resources Council where he is Assistant Director. He is presently serving as an Assistant Attorney General for the state of Iowa.

Service of the U.S. Department of Agriculture, total more than $7 billion in the construction of such works over the last thirty years.

Meanwhile, the average annual flood losses have increased from about $.5 billion to more than $1 billion. These huge sums do not include the costs of construction by other federal agencies of projects with incidental flood control benefits or the cost of construction of flood control works by state and local governments, agencies, private corporations, and individuals. And this is not to say that the flood control projects do not operate as designed or that there has been a hydrologic or hydraulic miscalculation. Rather, this amount represents steadily increasing developmental pressure on land along rivers and streams that are subject to inundation during floods.[4,18] This pattern demands consideration of alternatives or supplements to the time-honored plan of providing only for construction of engineering works to *control the river* in time of floods.[20,21] Obviously there is little benefit in reducing the flood peak on a given stream if the flood plain area below the protection works is subsequently developed at a correspondingly lower level or the downstream floodway is further encroached upon so as to materially reduce or eliminate the reduction in flood stages effected by the protective works.

This leads us to consideration of *direct control over land* adjacent to the river or stream as a means of flood damage minimization.[2,6,18,21] Assuming recognition of the need for this type of land-use control, many serious questions remain as to the type and extent of the regulations to be imposed and the level of government at which the control will be exercised.[3,5] The state of Iowa has concluded that basic responsibility for its flood plain management program should be centered at the state level.[14] This decision perhaps was based on the generally accepted concept of decentralization of control to the lowest level of government with the financial and jurisdictional capability to provide the control and attendant required services. This conclusion seems entirely appropriate since the federal government has no general police power and few Iowa municipalities and counties have the personnel required to perform the studies and formulate the controls needed in adequate flood plain management. Perhaps more important, floods generally have little respect for corporate limits or county boundaries. Too, there clearly seems to be a need in today's highly mobile society for the establishment of relatively uniform minimum protection levels and methods.[2,18]

The exercise of the state police power to regulate land use in the interest of flood damage prevention has been upheld by the U.S. Supreme Court and by the court of last resort in several states including California, Connecticut, Georgia, Missouri, New Hampshire, North Carolina, South Carolina, and West Virginia.[5,18]

## STATUTORY CONTROL OF FLOOD PLAINS IN IOWA

As is probably the case with most legislative enactments, the impetus for legislation in the area of water resources is usually provided by an extreme occurrence, either an abnormal shortage or a great surplus of water. In Iowa the disastrous floods of 1944 and 1947 prompted the appointment in 1948 of a legislative study committee to ascertain the needs and problems of the state with regard to its water resources and to recommend any needed legislation. The resulting legislation enacted in 1949 was the first comprehensive legislative recognition in Iowa of the interest of the state in its water resources.[7] This legislation was subsequently codified as Chapter 455A.[14] Other state laws, policies, and programs which apply to water resources and related land resources have been summarized and reported previously.[17,19]

### Establishment of the Iowa Natural Resources Council

The Iowa General Assembly in the 1949 act recognized that the protection of life and property from floods; the prevention of damage to lands therefrom; and the orderly development, wise use, and protection thereof is of paramount importance to the welfare and prosperity of the people of the state. Under the provisions of this act, the legislature created the Iowa Natural Resources Council (INRC) as the state agency to administer the act and vested in it the duty and authority to establish and enforce an appropriate comprehensive statewide program for the control, utilization, and protection of the surface water and groundwater resources of the state. Unfortunately, the policy statements of this original act were much broader than the specific powers granted, which were concerned primarily with floodway encroachment, flood control planning, and the regulation of after-constructed flood control works.[7] This initial law also assigned to the INRC the administration of the Iowa milldam law.

### Statutory Amendments of 1957 and 1965

The water shortages experienced during the mid-1950's prompted the Iowa legislature to enact a comprehensive water rights law in 1957 to regulate the use of surface and underground water in the state.

The act creating the water-use permit system also included provisions correcting some of the deficiencies in the original 1949 act in the area of floodway encroachment. In particular, the submission of applications and plans for floodway projects was made mandatory where the initial enactment apparently left this to the discretion of the developer.[8] This considerably enhanced the effectiveness of the statute and rapidly increased the work load of the INRC and its staff.[1]

The growing public awareness of the need for an alternative or supplement to construction of engineering works to control the river in time of flood prompted the 1965 Iowa legislature to strengthen further the water resources planning and regulatory provisions of Iowa law.[9,10,16] Amendments to the existing law provided specifically for the establishment and enforcement of comprehensive regulations for an entire stream or a significant reach thereof, without regard to whether a specific project was presently proposed for that area. The present organization of the activities and responsibilities of the INRC is shown in Figure 14.1.

## ADDITIONAL REGULATORY ACTIVITIES

Several statutes currently in force in Iowa, which authorize specified water-associated activities, also incidentally provide for control of those activities affecting or relating to flood control.[19] Generally, control is effected by providing in the statute for INRC approval of the flood control aspects of the authorized activity or project. For example, certain activities of the State Conservation Commission under Code Chapters 108, 109, 111, and 112; riverfront improvement commissions under Code Chapter 372; levee and drainage districts under Code Chapter 455; soil conservation districts under Code Chapter 467A; and soil conservation and flood control districts (which affect or relate to flood control) under Code Chapter 467C are subject to approval of the INRC. Approval of the council is also required for county and municipal zoning ordinances or regulations insofar as they relate to use of the floodway or flood plains.[11,12,13]

Direct control over the construction, operation, and maintenance of milldams is provided by Iowa Code Chapter 469.[15] A permit from the INRC is required under this statute for the construction of a dam on any meandered stream for any purpose or on any other stream for manufacturing or power purposes. The statute also provides for annual licensing and inspection of structures permitted thereunder.

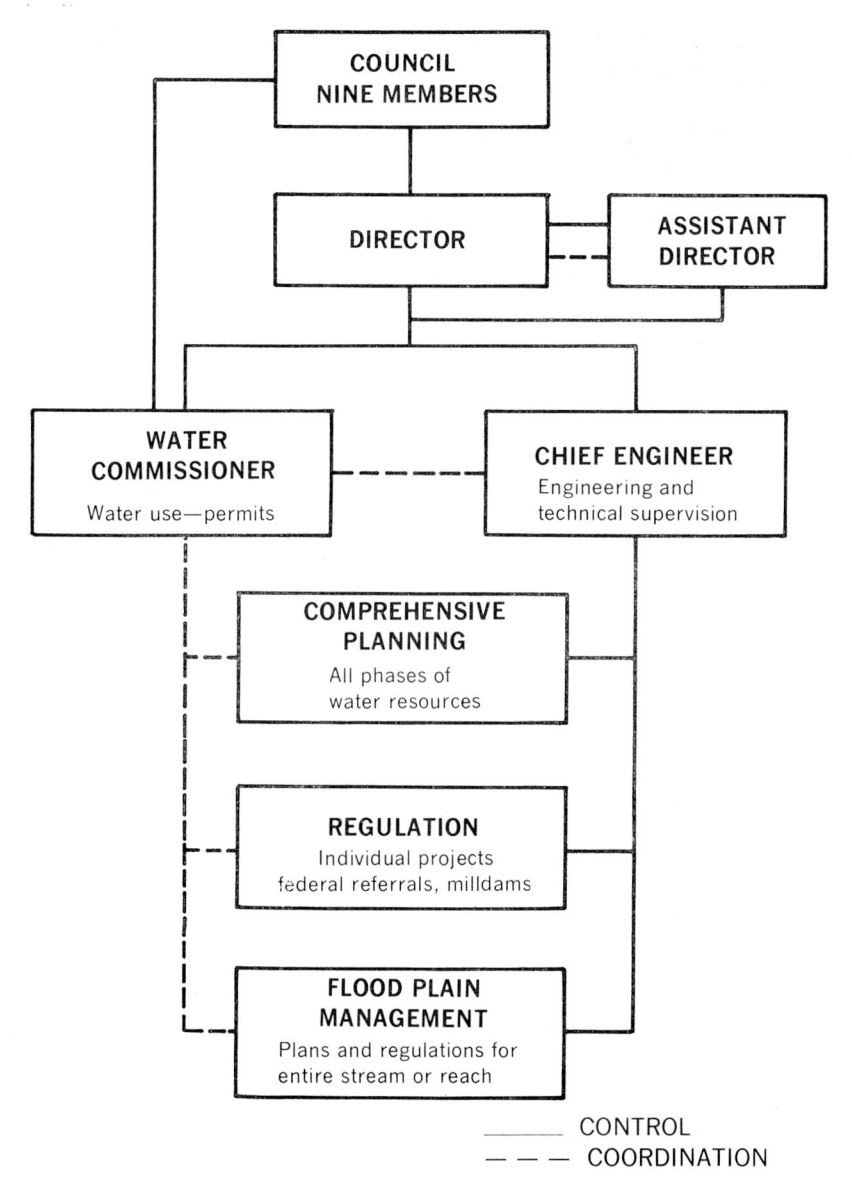

Fig. 14.1. Organizational chart for the Iowa Natural Resources Council, the state water resources agency.

The standards for the granting of such permits are contained in Section 469.5, as follows:

> 469.5 When permit granted. If it shall appear to the council that the construction, operation, or maintenance of the dam will not materially obstruct existing navigation or materially affect other public rights, will not endanger life or public health, and any water taken from the stream in connection with the project, excepting water taken by a municipality for distribution in its water mains, is returned thereto at the nearest practicable place without being materially diminished in quantity or polluted or rendered deleterious to fish life, it shall grant the permit, upon such terms and conditions as it may prescribe.

## PROVISIONS RELATING TO FLOOD PLAIN MANAGEMENT

General direct control over activities in or on floodways or flood plains is provided by Iowa Code Chapter 455A.[14] This chapter enunciates the broad policies of the state with regard to its water resources and vests the duty and authority for implementation of those policies in the INRC. Water occurring in any basin or watercourse or other natural body of water of the state is declared to be public water and public wealth of the people of the state, subject to use in accordance with the policies and principles of beneficial use set forth in the chapter. The water-use permit system is administered in accordance with these policy provisions and the specific powers enumerated in the statute.

### Comprehensive Water Resources Planning

Chapter 455A also authorizes the INRC to establish and enforce a comprehensive statewide plan for the control, utilization, and protection of the water resources of the state. The council is directed to coordinate state planning with local and national planning and to undertake the resolution of any conflicts that may arise between the water resources policies, plans, and projects of the federal government and those of the state, its agencies, and its people.

Pursuant to this authority, the INRC has completed preliminary studies of water needs and problems in the state and is now participating in comprehensive water resources planning efforts for the Upper Mississippi River basin and the Missouri River basin. All other state agencies and departments with an interest in specific areas of water resources, including the University of Iowa and Iowa State University, are also participating in this planning effort. Out of this coordi-

# STATE STATUTORY RESPONSIBILITY 163

nated effort will come the comprehensive statewide plan for water resources contemplated in the enactment of Chapter 455A.

## Jurisdictional Definitions

Under Chapter 455A the INRC is given the jurisdiction over the public and private waters in the state and the lands adjacent thereto necessary to carry out the provisions of the chapter and is authorized to construct flood control works or any part thereof. The statute contains very broad jurisdictional definitions as follows:

> "Flood plains" means the area adjoining the river or stream, which has been or may be hereafter covered by flood water; "Floodway" means the channel of a river or stream and those portions of the flood plains adjoining the channel, which are reasonably required to carry and discharge the flood water or flood flow of any river or stream; "Person" means any natural person, firm, partnership, association, corporation, state of Iowa, any agency of the state, municipal corporation, political subdivision of the state of Iowa, legal entity, drainage district, levee district, public body, or other district or units maintained or to be constructed by assessments, or the petitioners of a proceeding, pending in any court of the state affecting the subject matter of this chapter.

## Specific Regulatory Powers of the Resources Council

Specific powers to implement the sweeping policy statements relating to the protection of life and property from floods are set forth in three sections of Chapter 455A.

Section 455A.36 provides for control of construction, operation, and maintenance of all works of any nature for flood control in the state. Any person constructing or installing such works after the effective date of the statute must file with the INRC a verified written application therefor and obtain approval of the application, plans, and specifications for the proposed works. Approval or disapproval is based upon a determination of whether the proposed works in the plans and specifications will be in aid of and acceptable as part of or will adversely affect or interfere with flood control in the state; will adversely affect the control, development, protection, allocation, or utilization of the water resources of the state; or will adversely affect or interfere with the state comprehensive plan for water resources or an approved local water resources plan. This section applies to "all works" constructed by any "person" and to ". . . all drainage districts, soil conservation districts, projects undertaken by the state conservation commission, all public agencies including counties, cities, towns and all political subdivisions of the state and to all privately undertaken projects relating to or affecting flood control."

Section 455A.33 regulates miscellaneous projects on floodways or flood plains. Under this section any person desiring to erect, to make, or to suffer or permit a structure, dam, obstruction, deposit, or excavation (other than a dam constructed under authority of Chapter 469) to be erected, made, used, or maintained in or on any floodway or flood plains must file a verified written application therefor with the council, setting forth the material facts relating thereto. After investigation or hearing, the council enters an order permitting or prohibiting the project. Approval or disapproval is based upon a determination of whether the proposed project will ". . . adversely affect the efficiency of or unduly restrict the capacity of the floodway, adversely affect the control, development, protection, allocation, or utilization of the water resources of the state, or adversely affect or interfere with the state comprehensive plan for water resources, or an approved local water resources plan. . . ."

Section 455A.35 provides means of controlling development on the floodway or flood plains of an entire stream or reach thereof without regard to whether a specific project is presently proposed in the area to be controlled. Under this section the INRC is authorized to establish and enforce regulations for the orderly development and wise use of the flood plains of any river or stream in the state. The council is directed to determine the characteristics of floods which reasonably may be expected to occur and may by order establish appropriate encroachment limits, protection methods, and minimum protection levels. The order shall fix the length of flood plains to be regulated, shall fix the width of the zone between the encroachment limits so as to reserve sufficient floodway to carry and discharge flood flows, and shall fix the design discharge and water surface elevations for which protection must be provided. No such order may be issued without due notice thereof and public hearing thereon, with opportunity given for presentation of protests.

### State Coordination With Local Communities

The encroachment provisions of Section 455A.35 also provide for cooperation between the state and local communities in establishing and exercising this control of flood plain lands. The INRC may assist the local community in the establishment of encroachment limits, flood plain regulations, and zoning ordinances within the community. Any such regulations proposed by local units of government must be submitted to the council for review and approval prior to adoption by the local unit of government. Changes or variations from an approved ordinance or regulation as it relates to flood plain use also must have prior approval of the INRC. Upon approval of the local regulations, the construction of projects in conformity therewith do not require individual approval of the council. Primary responsibility for the administration and enforcement of the flood protection levels thereby established thus remains with the local community. This appears to be logical, desirable, and quite necessary under the existing manpower resources of the state agency.

## SUMMARY

It becomes obvious that the Iowa plan for flood plain management permits, encourages, and demands cooperation of all levels of government and the continuing cooperation of an informed citizenry. Amendments to the original organic act which established the INRC have extended and broadened this cooperative program. The council at the state level can establish uniform criteria for regulatory provisions and minimum protection levels for all flood plain areas. Within the general guidance of the state, the local counties and communities can plan cooperatively at the local or regional level for the optimum use of flood plain lands. Adoption of adequate flood plain regulations both by local agencies and the INRC will ensure that subsequent development conforms with comprehensive plans and that encroachments will not occur. The stage is now set for positive action, and the citizens of Iowa can be encouraged to support and participate in local programs for flood plain management.

## REFERENCES

1. Akin, W. E., and Dougal, M. D. Flood plain regulation in Iowa. In G. F. White, ed. *Papers on flood problems.* Dept. of Geog., Res. Paper 70. Chicago: Univ. of Chicago Press. 1961.
2. American Society of Civil Engineers. Guide for the development of flood plain regulations. Progress rept., Task Force on Flood Plain Regulations. *Proc. Am. Soc. Civil Eng., Hydraulics Div.* 88, no. HY5, Paper 3264 (Sept. 1962).
3. Beauchert, E. W. *A legal view of the flood plain.* Cambridge: Harvard Law School. 1961.
4. Cooter, H. H. To stay out of floods. *National Civic Review* 50, no. 10 (1961): 534–39.

5. Dunham, A. Flood control via the police power. *Univ. Pa. Law Rev.* 107 (1959): 1098–1132.
6. Iowa City, City of. Zoning ordinance 2238. (Article VI, flood plain use regulations.) July 1962.
7. Iowa, Acts. 53 G.A. Ch. 209. 1949.
8. Iowa, Acts. 57 G.A. Ch. 229. 1957.
9. Iowa, Acts. 61 G.A. Ch. 373. 1965.
10. Iowa, Acts. 61 G.A. Ch. 374. 1965.
11. Iowa Code. Ch. 358A. 1966.
12. Iowa Code. Ch. 368. 1966.
13. Iowa Code. Ch. 414. 1966.
14. Iowa Code. Ch. 455A. 1966.
15. Iowa Code. Ch. 469. 1966.
16. Iowa Natural Resources Council. *Report for the biennial period, July 1, 1962, to June 30, 1964*. Des Moines. 1964.
17. ———. *State laws, policies, and programs pertaining to water and related land resources in Iowa. Comprehensive Survey, Upper Mississippi River Basin*. Des Moines. 1967.
18. Murphy, F. C. *Regulating flood plain development*. Dept. of Geog., Res. Paper 56. Chicago: Univ. of Chicago Press. 1958.
19. Peterson, C. E. Iowa state laws, policies, and programs relating to water and related land resources. *Proc. Fourth Ann. Water Res. Design Conf.* Ames: Iowa State Univ. Eng. Ext. 1966.
20. Resources for the Future. *Neglected alternatives to flood protection*. Annual rept. Washington. 1964.
21. U.S. House of Representatives, Committee on Public Works. *A unified national program for managing flood losses*. Rept. of the Task Force on Federal Flood Control Policy. House Document 465. 89th Cong., 2nd sess. 1966.

# 15

## MODEL FLOOD PLAIN REGULATIONS FOR IOWA—A PROGRESS REPORT

### David J. Blair

FLOOD PLAIN REGULATION traditionally has been thought to involve physical control of watercourses. Channel improvements, levees, dikes, floodwalls, dams, and reservoirs have been constructed at enormous expense by all levels and units of government. The success of this effort to control flood damage through protective works is substantial in the affected localities, yet annual flood losses continue to rise. This circumstance, symptomatic of the ease and initial economy of human activity upon the flat borders of rivers and streams, compels land-use planners to seek alternative modes of control which are directed toward regulation of the flood plains themselves.

It is the purpose of this chapter to review and to report briefly on the legal and administrative problems and prospects of such regulation. This will provide the flood plain planner and/or manager with a good idea of the law concerning flood plain regulation. The Agricultural Law Center of the College of Law, University of Iowa,

DAVID J. BLAIR is a Law Research Assistant, Agricultural Law Center, College of Law, University of Iowa, where he is completing this flood plain legal research study under the guidance of N. Wm. Hines, Associate Professor of Law.

is concerned at present with research which will culminate in the formulation of model flood plain regulations for Iowa municipalities. The scope and content of this chapter represent a progress report on that effort.[1]

## THE LEGAL CONTEXT—CONSTITUTIONAL PROBLEMS

In the evolution of the law of real property, courts and legislatures have progressively carved exceptions into the common-law rule that a holder of realty in fee simple absolute (the highest estate in land) has absolute license to do on and with the land as he pleases. Thus courts have recognized that the owner may owe stringent duties of care to certain classes of persons temporarily using or visiting his land, and legislatures have announced that certain uses of land are so repugnant to private or public interests as to constitute legal nuisances. Indeed, these and similar exceptions so threaten to obscure the original rule that it now seems to retain little meaning, but cutting across the impact of such exceptions are the fundamental guarantees which reside in the United States Constitution and its state counterparts.

### Equal Protection

The equal protection concept[2] requires that statutory classifications be reasonably related to statutory purpose. Persons similarly situated must be similarly treated.[3] Assuming a valid purpose, dissimilar treatment is sustainable only if those affected are not similarly situated. This rule becomes important in flood plain regulation because persons situated upon flood plains inevitably must be subjected to restrictions, as compared to those residing on lands not subject to flooding.

Against the claim of flood plain property owners that such restrictions constitute a denial of equal protection, the proponent of the regulation must demonstrate that sound technical grounds exist for the differing classification. This presentation of compelling data is of highest importance. A showing that the persons affected are "generally" within inundation limits will be clearly insufficient. Rather, the proponent must show that, as to any claimant, reasoned data supports his individual inclusion in the restricted area.

## Due Process

Substantive[4] due process[5] is a broad requirement of reasonableness in legislation. In its various facets this guarantee demands that legislation (1) be designed to accomplish an end which the enacting authority has a right to achieve, (2) be reasonably calculated to achieve the legitimate end, and (3) be reasonable (not arbitrary or capricious) in relation to the legitimate end.[6] Thus for purposes of flood plain regulation, the due process concept will focus inquiry upon the allowable policy grounds of the regulation (the legitimate end), the probability that the regulatory scheme will achieve the allowable goal, and the reasonableness of the regulatory scheme. In regard to the latter, the question must be raised that although the means chosen will cure the flood hazard, are such means unreasonable in view of the kind of degree of hazard? These inquiries, too, may turn upon proper presentation of proof. Reasonableness is a function of hazard, and unusually restrictive regulations become reasonable upon a compelling demonstration of unusual hazard.[7]

Another constitutional guarantee generally included within the due process framework is that which prohibits the taking of private property for public purposes without adequate compensation being afforded the owner.[8] Here the inquiry is solely concerned with the effect of the regulation upon private property. If the regulation is so restrictive as to be confiscatory in effect (a "taking"), courts will hold that the governing body must pay its own way through condemnation proceedings.[9] Flood plain regulation, of course, cannot achieve its goals without proscribing many uses of property. It is thus important for constitutional validity that the regulation include more than merely negative proscriptions, and probable validity will be strengthened by the inclusion of broad categories of permitted uses. Each such category weakens the claimant's argument that his property, in practical effect, has been "taken" without compensation.

## THE LEGAL CONTEXT—STATE ENABLING LEGISLATION

Under our federal system of government, those powers not delegated to the national government are retained by the states. These retained powers have come to be known cumulatively as the "police power." The state under its police power is competent to legislate broadly for public health, safety, morals, and general welfare.

Municipal corporations are creatures of the state, existing solely at the pleasure of the state, and capable of exercising only those powers which the state has delegated to be so used.[10] The question initially arising, then, upon the exercise of legislative power by a municipality is whether an appropriate delegation of that power has been made by the state.

For purposes of flood plain regulation, Iowa municipalities may look to three important sources of delegated police power.

First, municipal corporations have been delegated a broad power of condemnation of land for public purposes.[11] Certainly in the floodway (the existing channel and such greater portion of land as is required to convey floodwaters of a given magnitude) and probably in the flood plain (those additional portions of land not required for the conveyance of floodwater but subject to inundation from floods of given magnitude), the requisite public purpose can be shown. Encroachments in the flood channel, insofar as they unreasonably raise upstream water surface elevations to the detriment of upstream property, would seem subject to condemnation on the most unenlightened reading of the public-purpose requirement. Property in the flood plain, although ineffectual upon floodflow by definition, can similarly be brought within the public-purpose rule by citing a detrimental effect upon health, safety, or general welfare. Condemnation is thus a possible tool in the planning of flood hazard areas. Its great disadvantages are cost and complete inflexibility of operation.

Second, municipal corporations have been delegated general power to enact ordinances "to provide for the safety, preserve the health, promote the prosperity, improve the morals, order, comfort, and convenience of such corporations and the inhabitants thereof. . . ."[12] This enabling statute makes no mention of the "general welfare" as a permissible policy ground of municipal ordinances, but the Iowa court has held that the state police power, of which general welfare is a part, is thereunder conferred.[13] It should also be noted that the specifically designated policy grounds—safety, health, prosperity, morals, order, comfort, convenience—constitute particular applications of a delegated power which by inference approaches in extent the power to legislate for general welfare.[14] Thus ordinances passed pursuant to this grant need not be limited to the typical "general safety" purposes. Clear statutory language gainsays such a restrictive reading. Municipal power to enact ordinances for the stated purposes clearly furnishes another source of power for flood plain regulation.

# MODEL FLOOD PLAIN REGULATIONS FOR IOWA

Finally, municipal corporations have been delegated the zoning power "for the purpose of promoting the health, safety, morals, or the general welfare of the community. . . ."[15] The statutory "general welfare" language leaves little doubt regarding the scope of this grant, and the Iowa court has held that it amounts to a delegation of the state police power.[16] There seems to be no ground for distinguishing the flood plain from any other zoning district; and flood plain regulations enacted under this delegation, if consonant with a comprehensive zoning plan, should be upheld if otherwise valid.[17]

## THE POLICY OF FLOOD PLAIN REGULATION

The preceding discussion suggests that, in addition to the condemnation technique, there exist two sources of delegated legislative power upon which municipal flood plain regulation might be based. Before the relative merits of these sources—the "general safety" and "zoning" powers—are evaluated, however, the draftsman must articulate the policy grounds supportive of his regulations. These policies fall under the general headings of preserving health, promoting safety, and avoiding economic loss.

### Health Problems

Disease follows floodwater with demonstrable certainty. The community disruptions incident to flooding—probable contamination of water supply, lack of sanitation, contamination of foodstuffs, improper disposal of waste materials, and lack of normal individual hygienic measures made impossible through circumstance—give rise to a wide range of infectious disease. Such diseases are primarily viral, bacterial, and mycotic infections, including but not limited to typhoid fever, paratyphoid diseases, tetanus, and skin infections due to uncleanly conditions.

These health hazards are compounded by the normal community response to flooding. Thus not only are persons residing in the flood plain subject to flood-associated diseases but also vulnerable are the many persons temporarily working there—rescue and relief crews, municipal employees, and the inevitable scores of volunteer flood workers. This clearly generalized threat to community health is a valid subject of community concern. As such, the preservation of health constitutes a sound policy of flood plain regulation.

### Concern for Community Safety

Municipalities may properly wish to promote the safety of their inhabitants. Again, floods generate overt physical dangers not only to flood plain residential and industrial personnel but also to the temporary personnel necessitated by flood conditions. These hazards range from death, proximately caused by floodwaters, to the proportionately larger numbers of injuries incident to vigorous physical activity generally and caused by flooding only indirectly. It seems that the prevention of this entire spectrum of potential physical harm could be the subject of community concern. As such, safety is a likely policy ground of regulation.

### Impact of Economic Losses

Beyond property damage losses within flooded areas, floods cause two varieties of economic loss of general community impact. These have an important bearing on regulatory aspects.

First, municipalities bear the large burden of funding local efforts to avoid or minimize flood hazards which threaten health, safety, and property. Municipal employees often function in rescue, sandbagging, and cleanup operations; and city funds may subsidize living expenses of flood victims until normal conditions return.

Second, and perhaps of greater concern, flood conditions bring the economic life of the community to an abrupt halt. As local efforts focus upon prevention of flood damage, merchants do not sell and consumers do not buy. The normal commercial ties with outside areas are weakened, for neither salesmen nor nonresident consumers wish to visit flood hazard areas.

Municipalities may reasonably regard the avoidance of such generalized economic loss as a desirable goal of legislation, and this goal furnishes a final policy ground of flood plain regulation.

## ALTERNATIVE MODES OF FLOOD PLAIN REGULATION —STATE ENABLING LEGISLATION

The inquiry now becomes the extent to which the above policy grounds—health, safety, economic loss—are legitimate goals of municipal legislative action. It is suggested not only that such goals are permissible within constitutional guidelines but also that alternative sources of delegated state power exist, upon which flood plain regulations (looking to these goals) might be based.

## Regulation by General Safety Ordinance

Flood plain regulation traditionally has been effected under the zoning power.[18] For many Iowa municipalities, however, this mode of control may be impossible. Any zoning ordinance must be adopted pursuant to a comprehensive zoning plan.[19] In the absence of such a comprehensive plan, which to this point substantial numbers of Iowa municipalities have found unadvisable or unnecessary,[20] flood plains cannot be regulated through zoning. Thus in these communities, and perhaps in others where zoning ordinances are political liabilities, it becomes desirable to seek alternative forms of regulation.

As noted above, municipal authority to enact "general safety" ordinances (Chapter 366) furnishes a likely source of legislative power.

Ordinances generally must be reasonably certain and specific,[21] consistent with state law,[22] and limited to a single subject.[23] If otherwise valid, they possess the force and effect of state law within the territorial bounds of the municipality.[24] Ordinances are not to be confined to the prevention of common-law nuisances.[25]

Favoring use of Chapter 366 for purposes here discussed, the Iowa court has held that ordinances are entitled to a strong presumption of reasonableness,[26] that the burden of proof is on the party attacking an ordinance,[27] that construing an ordinance is a question of law for the court,[28] that a city cannot be enjoined from initially enacting an ordinance,[29] and that the court has neither the power nor the right to inquire into the wisdom or sound judgment of ordinances.[30] Also of great interest for present purposes, the Iowa court has held that there can be no valid objection to a police power regulation on the basis that it is made applicable to a segregated area or district.[31]

The permissible policy grounds of Chapter 366 ordinances are safety, health, prosperity, morals, order, comfort, and convenience.[32] It has been held that this grant amounts to a delegation of the state police power to be exercised in the interest of public welfare,[33] and it seems that such a grant sufficiently encompasses the policies of flood plain regulation (health, safety, economic loss). The health and safety grounds are, of course, specifically mentioned in the statute as goals which municipalities have a right to achieve. Similarly, the aspect of flood plain regulation aimed at minimizing economic loss is echoed in the statutory authorization of community "prosperity" as a permissible goal. And, finally, insofar as floods tend to disrupt community order, comfort, and convenience, it seems that the elimination of such evils through flood plain regulation finds firm support in the statute.

Until recent years the rule obtained in Iowa that delegations of the police power to municipalities were to be strictly construed.[34] Such a rule suggested that in a close case, wherein the issue involved the existence of municipal power to reach a questionable goal, a court might well hold the goal to be outside the bounds of delegated power. In a 1963 amendment to Chapter 368 (General Powers of Municipal Corporations), however, the legislature expressed its clear disapproval of the strict-construction doctrine. The important language of this rule of construction amendment is worthy of quotation:

> Cities and towns are bodies politic and corporate . . . and shall have the general powers and privileges granted, and such others as are incident to municipal corporations of like character, not inconsistent with the statutes of the state, for the protection of their property and inhabitants, and the preservation of peace and good order therein. . . .
>
> It is hereby declared to be the policy of the state of Iowa that the provisions of the Code relating to the powers, privileges, and immunities of cities and towns are intended to confer broad powers of self-determination as to strictly local and internal affairs upon such municipal corporations and should be liberally construed in favor of such corporations. The rule that cities and towns have only those powers expressly conferred by statute has no application to this Code. Its provisions shall be construed to confer upon such corporations broad and implied power over all local and internal affairs which may exist within constitutional limits. . . .[35]

Clearly, the strict-construction doctrine has been legislatively disapproved, and there seems to be no present indication that courts have narrowed by their interpretation the scope of this amendment.[36] For our purposes, however, it is essential to note that the amendment is literally a rule of construction. No new substantive powers are granted. Rather, the "provisions of the Code relating to the powers . . . of cities and towns" are to be "liberally construed" in favor of "broad and implied power over all local and internal affairs which may exist within constitutional limits."[37] This then is further ground for asserting that municipalities, as recipients of a broad delegation of state police power, may enact general safety ordinances for flood plain regulation. Indeed, it seems that a court would be unable to find such a goal invalid on grounds of exceeding the delegated power. The power to enact general safety ordinances is a sound source of flood plain regulatory authority.

## Regulation by Zoning Ordinance

Alternatively, municipalities might reasonably choose to regulate flood plains through the zoning power.[38] Such a choice allows the use of preexisting administrative bodies—the plan and zoning commission and board of adjustment—and familiar modes of procedure. Among other advantages, there is some indication that courts construe the zoning power with liberality. Thus the Iowa court has said that a city council enjoys wide discretion in the framing of zoning ordinances[39] which are presumptively reasonable and valid.[40] It is not ground for complaint that an uncompensated burden is placed upon some property.[41] Zoning regulations, in recognition of the principle that private property may be subjected to restrictions for the good of the community,[42] may adversely affect the market value of property.[43] Such adverse effect may be considered by the court, but lessened market value will not be determinative of validity.[44] The prime consideration is not the hardship of the individual case, but the general purpose of the ordinance and its relationship to that purpose.[45] The burden of proof is on the party attacking the ordinance.[46]

The Iowa Code provides that for zoning purposes a municipality may be divided into districts[47] and that zoning regulations may be prescribed for each district so established.[48]

As to regulations within districts, we need not even ask if the policy grounds of flood plain zoning are rendered permissible by the statute. The legislature in a 1965 amendment[49] added "safety from . . . floods" to the list of valid regulatory purposes, and it is clear that regulations within districts may be aimed at minimizing flood hazard.

With regard to establishing districts, however, the Code provides:

> For the purpose of promoting the health, safety, morals, or the general welfare of the community, any city or town is hereby empowered to regulate . . . buildings and other structures. . . .[50]

The Code section on "Districts" provides:

> For any or all of said purposes the local legislative body . . . may divide the city or town into districts of such number, shape, and area as may be deemed best suited to carry out the purposes of this chapter. . . .[51]

We are given two references by the latter section. Initially, we are told that zoning districts may be established for purposes of health, safety, and general welfare (former section); but we are also told that the "number, shape, and area" of the districts are to be established with regard to "the purposes of this chapter." This latter language at least arguably refers us to the section on regulations within districts, wherein we have seen that minimizing flood hazard is a clearly permissible policy ground. But even assuming that a court would not so hold, it seems that the grounds of flood plain zoning—health, safety, economic loss—may be fitted within the more general language of the statute. The health and safety grounds are, of course, repeated in the statute. The only question is whether general welfare includes avoidance of economic loss. Three arguments suggest an affirmative answer:

1. The city plan commission is charged under Code procedure with developing a comprehensive zoning plan. The plan should be designed to "promote health, safety, . . . prosperity, and general welfare, as well as efficiency and economy in the process of development."[52] It is absurd to suppose that, although the plan commission should develop its plan with regard to economic factors, the city council in enacting the same plan should disregard economic factors. Therefore, a reasonable reading of the Code requires the inclusion of avoidance of economic loss within the catchall general welfare term.
2. It has been held that Chapter 414 (Municipal Zoning) amounts to a delegation of the state police power.[53] Chapter 366 (Ordinances), which lists "prosperity" as a policy ground, has been similarly construed.[54] In the absence of compelling indications that the police power should be inconsistently construed, its delegation under Chapter 414 should include "prosperity" as a policy ground of zoning.
3. It has been held that the policy grounds of Chapter 366, which include prosperity, amount to the "general welfare."[55] Chapter 414 lists general welfare as a policy ground. Therefore, the latter version of general welfare should be consistently construed to include prosperity.

It is suggested that these arguments, cumulatively if not individually, adequately support the position that municipalities may enact flood plain zoning ordinances for all the purposes of health, safety, and avoidance of economic loss. In due process terms, they are all ends which the municipality has a right to achieve. This conclusion is

**MODEL FLOOD PLAIN REGULATIONS FOR IOWA** 177

further supported by the rule of construction amendment,[56] which cautions courts to construe municipal powers with liberality as to local affairs.

## Zoning or General Safety Ordinance

The preceding discussion suggests that flood plain regulation may be accomplished either through traditional zoning techniques or through general municipal power to enact ordinances. Chapter 455A of the Iowa Code recognizes this possibility:

> The council [Iowa Natural Resources Council] may cooperate with and assist local units of government in the establishment of . . . flood plain regulations and zoning ordinances relating to flood plain areas. . . . Encroachment limits, flood plain regulations, or flood plain zoning ordinances . . . shall be submitted to the council for review. . . . Changes or variations from an approved regulation or ordinance as it related to flood plain use shall be approved by the council prior to adoption.[57]

The way is thus clear in Iowa for flood plain regulation, either through zoning ordinances or general safety ordinances. The choice of legislative vehicle should turn upon considerations other than validity, for both are equally valid and both may be employed to reach all the policy grounds of flood plain regulation.

## MODEL FLOOD PLAIN REGULATIONS

As stated at the outset, the purpose of this chapter is to report on the task of developing model regulations.[58] The substantive form which the regulations should take are not yet established, but several provisions certainly should be included.

### Two-Zone Technique

It seems that two flood districts should be established, the floodway and the flood plain. The *floodway* is defined as the present channel plus those portions of land which are needed to convey a flood of designated magnitude without unreasonably raising upstream water surface elevations. The *flood plain* (termed floodway fringe by some) is defined as those additional portions of land not required for conveyance of the flood of designated magnitude but subject to inundation.[59] The precise boundaries of these districts should be determined by the Iowa Natural Resources Council, acting upon the best available hydrologic data in cooperation with the local city council.[60]

Structures in the flood channel, because of their demonstrable effect upon water elevations, constitute relatively urgent subjects for elimination. It thus seems that these structures should be condemned and removed by the municipality.

In the flood plain, however, it seems that regulations should be established looking to reasonable private uses of the affected property. If a comprehensive zoning plan already exists, the new regulations should be superimposed upon existing classifications.

As to future development of the flood plain, private owners should be presented with two possibilities. First, by landfill construction they may raise the level of property above projected floodwater elevations, thus becoming eligible for any use otherwise permitted within the district. Alternatively, a reasonably exhaustive listing of uses permitted without landfilling should be made. This category of uses will be developed in accordance with the policy grounds of the ordinance—health, safety, economic loss—and insofar as possible will not be limited solely to open uses.

**Nonconforming Uses**

Draftsmen of model regulations have often excluded any provision dealing with nonconforming uses. The rationale of this approach is (1) that nonconforming uses are more conveniently handled in the community's land-use regulations of general application and (2) that the nonconforming use issue raises political opposition which may foreclose the enactment of any flood plain regulations. It is suggested that these objections are unsound. First, the reasonableness of restrictions on property is a function of hazard and necessity. The strong policy grounds of flood plain regulation require a nonconforming use provision which is tailored to the statutory goals. Second, control of preexisting uses is an urgent goal of regulation. Future development of the flood plain is of higher importance, but out of existing uses arise the most presently grave threats to community health, safety, and economy. If the nonconforming use provision generates political opposition, then that battle is nonetheless preferable to flood plain regulation which is an empty letter.

The nonconforming use section should be a composite of a large number of provisions, including:

# MODEL FLOOD PLAIN REGULATIONS FOR IOWA

1. Restrictions against structural alteration, extension, and enlargement.
2. Restrictions against lapse of the nonconforming use.
3. Restrictions against repair if the structure is, for example, 50 percent destroyed.
4. Restrictions based on the projected useful life of the building at the time the ordinance is enacted.
5. An ultimate restriction that any nonconforming use shall not be permitted after, for example, 50–75 years from the time the ordinance is enacted.

It is believed that a reasonably restrictive nonconforming use provision, in combination with an extensive listing of permitted uses, will withstand constitutional challenge. The most serious challenge would be on the ground that private property has been taken for a public use without just compensation. Such a provision for nonconforming uses, the essence of flood plain control, will be included in the model regulations.

## Regulating One of Several Watercourses

An additional problem arises when a municipality seeks to regulate one of several flood plains within its corporate limits. This approach, singular but cumulative with time, may be desirable when the burden of gathering technical data for all watercourses would unreasonably delay implementation of any flood plain regulations. There immediately arises, however, the question of constitutional objections.

No due process objections should succeed. Under the familiar principle that legislation may be framed to cure the worst facet of the problem and that not all the problem need be initially cured,[61] it seems that the flood plain of the most menacing watercourse might constitutionally be regulated first.

Nor should challenges on equal protection grounds be successful. The claimant would assert that the regulation, insofar as it applied to him and not to others similarly situated, constituted a denial of equal protection. It suffices to answer that the claimant and others are not similarly situated. Because the claimant is situated on the most dangerous flood plain, a reasonable city council could conclude that his property should receive a different classification than that situated in a less dangerous flood area.

## CONCLUSIONS

Despite the construction of ambitious protective works, annual flood losses continue to rise; and flood plain regulation—directed toward control of property and structures situated within flood plains—becomes an important mode of municipal land-use planning. The goals of flood plain regulation—health, safety, and avoidance of economic loss—may be reached constitutionally under state enabling legislation through either the zoning or general safety powers. Model regulations, drafted in alternative form to recognize this possibility, are presently being prepared. When completed and published, they should make easier the task of the flood plain manager and others concerned with comprehensive flood plain management.

## REFERENCES

1. The *Model Flood Plain Regulations* will be published upon completion of current studies. Requests for information, reports, or reprints should be directed to the Agricultural Law Center, College of Law, University of Iowa, Iowa City, Iowa. Also, consistent with the nature of this progress report, many footnotes are omitted.
2. Iowa Const., art. I, § 9.
3. See *Keller* v. *City of Council Bluffs*, 246 Iowa 202, 66 N.W.2d 113 (1954); *City of Sioux City* v. *Simmons Hardware Co.*, 151 Iowa 334, 131 N.W. 17 (1911).
4. This is to be distinguished from procedural due process, a concept which requires that the *channels* through which the substantive rights and liabilities of the parties are adjusted be reasonable and fair.
5. Iowa Const., art. I, § 9.
6. See *Central States Theatre Corp.* v. *Sar*, 245 Iowa 1254, 66 N.W.2d 450 (1954); *Granger* v. *Board of Adjustment*, 241 Iowa 1356, 44 N.W.2d 399 (1950); *Anderson* v. *Jester*, 206 Iowa 452, 221 N.W. 354 (1928).
7. See *Brackett* v. *City of Des Moines*, 246 Iowa 249, 67 N.W.2d 542 (1954).
8. Iowa Const., art. I, § 18.
9. See *Keller* v. *City of Council Bluffs*, 246 Iowa 2021, 66 N.W.2d 113 (1954).
10. See Iowa Code, § 366.1 (1966).
11. See Iowa Code, § 368.37 (1966).
12. See Iowa Code, § 366.1 (1966).
13. See *Town of Grundy Center* v. *Marion*, 231 Iowa 425, 1 N.W.2d 677 (1942); *Harris* v. *City of Des Moines*, 202 Iowa 53, 209 N.W. 454 (1926); *City of Des Moines* v. *Manhattan Oil Co.*, 193 Iowa 1096, 184 N.W. 823 (1922).
14. See *City of Osceola* v. *Blair*, 231 Iowa 770, 2 N.W.2d 83 (1942).
15. See Iowa Code, § 414.1 (1966).
16. See *City of Bloomfield* v. *Davis County Com. Sch. Dist.*, 254 Iowa 900; 119 N.W.2d 909 (1963); *Granger* v. *Board of Adjustment*, 241 Iowa 1356, 44 N.W.2d 399 (1950); *Boardman* v. *Davis*, 231 Iowa 1227, 3 N.W.2d 608 (1942).

## MODEL FLOOD PLAIN REGULATIONS FOR IOWA

17. See *Downey* v. *City of Sioux City*, 208 Iowa 1273, 227 N.W. 125 (1929).
18. See Iowa Code, ch. 414 (1966).
19. See *Downey* v. *City of Sioux City*, 208 Iowa 1273, 227 N.W. 125 (1929).
20. Precise figures on comprehensive zoning by Iowa municipalities are unavailable at this writing. Tentative results of a survey of Iowa counties indicate, however, that only a handful have adopted comprehensive zoning ordinances.
21. See *Edwards & Browne Coal Co.* v. *City of Sioux City*, 213 Iowa 1027, 240 N.W. 711 (1932).
22. See Iowa Code, § 366.1 (1966).
23. See Iowa Code, § 366.2 (1966).
24. See *Jurgens* v. *Davenport, R.I. & N.W. Ry.*, 249 Iowa 711, 88 N.W.2d 797 (1958); *Boardman* v. *Davis*, 231 Iowa 1227, 3 N.W.2d 608 (1942).
25. See *Town of Grundy Center* v. *Marion*, 231 Iowa 425, 1 N.W.2d 677 (1942).
26. See *Iowa City* v. *Glassman*, 155 Iowa 671, 136 N.W. 899 (1912).
27. See *Jurgens* v. *Davenport, R.I. & N.W. Ry.*, 249 Iowa 711, 88 N.W.2d 797 (1958); *Snouffer* v. *Cedar Rapids & M.C. Ry.*, 118 Iowa 287, 92 N.W. 79 (1902). But see Iowa Code, § 455A.37 (1966).
28. See *City of Creston* v. *Mezvinsky*, 213 Iowa 1212, 240 N.W. 676 (1932); *Platt & Speith* v. *Chicago, B. & Q. Ry.*, 74 Iowa 27, 37 N.W. 107 (1888).
29. *Des Moines Gas Co.* v. *City of Des Moines*, 44 Iowa 505 (1876).
30. See *Hirsch* v. *City of Muscatine*, 233 Iowa 590, 10 N.W.2d 71 (1943); *Scott* v. *City of Waterloo*, 223 Iowa 1169, 274 N.W. 897 (1937); *Edaburn* v. *City of Creston*, 199 Iowa 669, 202 N.W. 580 (1925).
31. See *City of Des Moines* v. *Manhattan Oil Co.*, 193 Iowa 1096, 184 N.W. 823 (1922).
32. See Iowa Code, § 366.1 (1966).
33. See, e.g., *City of Osceola* v. *Blair*, 231 Iowa 770, 2 N.W.2d 83 (1942); *Town of Grundy Center* v. *Marion*, 231 Iowa 425, 1 N.W.2d 677 (1942); *Cecil* v. *Toenjes*, 210 Iowa 407, 228 N.W. 874 (1930).
34. See, e.g., *Dotson* v. *City of Ames*, 251 Iowa 467, 101 N.W.2d 711 (1960); *Downey* v. *City of Sioux City*, 208 Iowa 1273, 227 N.W. 125 (1929); *Ebert* v. *Short*, 199 Iowa 147, 201 N.W. 793 (1925).
35. See Iowa Code, § 368.2 (1966).
36. See *Richardson* v. *City of Jefferson*, 257 Iowa 709, 134 N.W.2d 528 (1965).
37. See Iowa Code, § 368.2 (1966).
38. See Iowa Code, ch. 414 (1966).
39. See *Herman* v. *City of Des Moines*, 250 Iowa 1281, 97 N.W.2d 893 (1959); *Keller* v. *City of Council Bluffs*, 246 Iowa 202, 66 N.W.2d 113 (1954).
40. Ibid.
41. See *Boardman* v. *Davis*, 231 Iowa 1227, 3 N.W.2d 608 (1942).
42. See *Call Bond & Mortgage Co.* v. *City of Sioux City*, 219 Iowa 572, 259 N.W. 33 (1935).
43. See *Boardman* v. *Davis*, 231 Iowa 1227, 3 N.W.2d 608 (1942); *Anderson* v. *Jester*, 206 Iowa 452, 221 N.W. 354 (1928).
44. See *Plaza Recreational Center* v. *City of Sioux City*, 253 Iowa 246, 111 N.W.2d 758 (1961); *Brackett* v. *City of Des Moines*, 246 Iowa 249, 67 N.W.2d 542 (1954).
45. See *Brackett* v. *City of Des Moines*, 246 Iowa 249, 67 N.W.2d 542 (1954).
46. See *Plaza Recreational Center* v. *City of Sioux City*, 253 Iowa 246, 111 N.W.2d 758 (1961); *Keller* v. *City of Council Bluffs*, 246 Iowa 202, 66

N.W.2d 113 (1954). But for the possibility of shifting the burden of proof, see Iowa Code, § 455A.37 (1966).
47. See Iowa Code, § 414.2 (1966).
48. Ibid.
49. See Iowa Code, § 414.3 (1966).
50. See Iowa Code, § 414.1 (1966).
51. See Iowa Code, § 414.2 (1966).
52. See Iowa Code, § 373.18 (1966).
53. See *City of Bloomfield* v. *Davis County Com. Sch. Dist.*, 254 Iowa 900, 119 N.W.2d 909 (1963); *Boardman* v. *Davis,* 231 Iowa 1227, 3 N.W.2d 608 (1942); *Anderson* v. *Jester,* 206 Iowa 452, 221 N.W. 354 (1928).
54. See *Cecil* v. *Toenjes,* 210 Iowa 407, 228 N.W. 874 (1930); *Harris* v. *City of Des Moines,* 202 Iowa 53, 209 N.W. 454 (1926); *City of Des Moines* v. *Manhattan Oil Co.,* 193 Iowa 1096, 184 N.W. 823 (1922).
55. See *City of Osceola* v. *Blair,* 231 Iowa 770, 2 N.W.2d 83 (1942).
56. See Iowa Code, § 368.2 (1966).
57. See Iowa Code, § 455A.35 (1966).
58. An attempt is being made to survey the present status of national flood plain regulation. State water agencies have generally been extremely helpful in furnishing current information regarding state and local legislation and recent court decisions. Local regulation, in the few municipalities wherein adopted, seems to be effected mainly through the zoning technique, with occasional use of building codes and subdivision regulations. Full documentation of regulatory patterns will appear in the final report.
59. These definitions substantially follow the statutory language, see Iowa Code, § 455A.1 (1966).
60. This procedure is authorized by statute, see Iowa Code, § 455A.35 (1966).
61. See *Dickinson* v. *Porter,* 240 Iowa 393, 35 N.W.2d 66 (1948); *Merchants Supply Co.* v. *Iowa Employment Security Commission,* 235 Iowa 372, 16 N.W.2d 572 (1944).

# 16

## BUILDING CODES AND LOCAL ZONING AND SUBDIVISION CONTROL FOR FLOOD PLAIN MANAGEMENT

### Jim L. Maynard

CAN INDIVIDUAL PROPERTY OWNERS and local agencies, on their own initiative, evaluate the flood risk and determine the precautions necessary to minimize their flood losses? Perhaps they could if they possessed the initiative and perseverance of a local citizen of Austin, Texas, who visited the office of this author at that city in the early 1960's. He inquired as to the availability of topographic maps of Austin and surrounding areas and was particularly concerned about elevation data for his residence and several other higher areas in that part of the city. Contour maps of several scales and contour intervals were made available to him, and he made careful notes of a number of elevations in a directional pattern from his home. When asked if additional information was desired, he disclosed his purpose, "I just wanted to see how high I should make my TV antenna to get San Antonio!"

JIM L. MAYNARD, A.I.P., is Director of Planning, Powers-Willis and Associates, Planners-Engineers-Architects, Iowa City, Iowa.

Now if everyone occupying flood plain lands and building near a river, stream, or watercourse would be as concerned about the location and elevation of their structures in relation to the flood plain as our Austin citizen was about the height of his television antenna, it perhaps would not be necessary for our local governments to concern themselves with exercising controls over the flood plain areas within their jurisdiction. Unfortunately this is not the case, as evidenced by these figures presented by S. Kenneth Love, Chief of the Quality of Water Branch, U.S. Dept. of the Interior, at the 1963 Conference of the American Society of Planning Officials:[9]

> In 1900, (annual) flood damage in the United States was about $100 million; in 1960 it was about $300 million. The increase is not due to a greater number of floods but to increased encroachment on flood plains. It has been estimated that for every six dollars spent by the federal government on flood protection, five dollars is spent by the general public expanding onto the flood plains. This expansion results from (1) ignorance that the area is subject to flooding, (2) failure of developers to warn prospective buyers of land that may be flooded, (3) the tendency of people to prefer living and working on level bottom lands, and (4) the higher value of flood plains and hence the source of higher tax revenues (in many hilly and mountainous areas).

Current estimates of annual expenditures for flood control projects purportedly exceed $500 million, and annual flood losses are approaching $1 billion.[13] Thus it is obvious that urban growth has not taken cognizance of the flood hazard. Additional and comprehensive planning and regulatory measures are needed.

It is therefore the purpose of this chapter to elaborate on a number of the regulatory measures that county, city, and town governments may adopt to guide and control the development of land areas subject to periodic inundation. These include zoning regulations, subdivision regulations, building codes, housing codes, and miscellaneous regulatory controls. Administration is an additional factor that will be included in this discussion.

## PRELIMINARY CONSIDERATIONS

Three principal points need to be emphasized in regard to flood plain regulations. Mention was made previously of the need for both county and municipal regulation, and this combined emphasis is important. Though we usually think of exercising control only over urban land uses such as residential subdivisions, commercial developments, and industrial complexes, there is cause for equal concern over

## BUILDING CODES, LOCAL ZONING, SUBDIVISION CONTROL 185

the location of rural developments. These include farmsteads, nonfarm residences, vacation resorts, summer cottages, private recreation developments, extraction of timber and natural resources, and other similar activities of an intensive, nonagricultural type that occur in the unincorporated areas of our counties. It should also be noted that in both land area and length of valley the greatest portion of our flood plains are under the jurisdiction of our county boards of supervisors.

Another point to keep in mind is that any watercourse, regardless of its present condition or past history, can become a threat to flood plain development if encroachments downstream obstruct the flow of flood waters or if the slopes of the watershed are urbanized to a degree that the runoff during periods of intensive or extended rainfall is doubled or tripled. It is therefore essential that planners and engineers remain aware of the existing capacity of watercourses and of the effect any sizable or strategically located development may have on that watercourse.[1,10]

Finally, it is suggested that our primary concern in regulating within the concept of the public interest ought to be for the ultimate occupant or owner of the land. Those through whose hands it may pass in the course of its conversion from raw or agricultural land to developed land for residential, industrial, or commercial uses seem to be able to take care of their immediate interests, though some guidance along the way can often be beneficial to them.[8] The long-range community interest, however, is that of protecting the health, safety, and general welfare of the individual occupant and owner. He is the one who must pay taxes and maintain the developed land. He is the one who expects to be served with utilities, streets, snowplows, school buses, fire trucks, garbage pickup, and police protection. And he is the one we must bail out when flood disaster strikes. Thus, the more we do to prevent conditions that could be detrimental to the ultimate occupant or owner, the more we will benefit as a community, physically and socially as well as economically.

In addition to the various levels of planning discussed in previous chapters, we have at our disposal a number of legislative tools which may be utilized by county boards of supervisors and city and town councils to regulate the development and use of land for various purposes. Among these are zoning ordinances, building codes, minimum housing standard codes, subdivision regulations, purchase or easements and development rights, and restrictive covenants. Through these regulations, comprehensive plans for guiding community growth are effectuated. The basic aspects of each regulatory measure and its specific application to regulation of the use of the flood plain will be studied in the following sections.

## ZONING REGULATIONS

In Iowa both municipalities and counties have been granted authority to enact zoning regulations. Chapter 414 of the Iowa Code has made this power available to cities and towns since 1924. It was passed by the legislature, according to a news report, "without amendment or dissenting vote." County zoning first became available in 1947 but was limited to larger counties and required a referendum. In 1955, Chapter 358A was amended to make zoning available to all counties, and in 1959 the referendum requirement was removed from the act, placing the decision on adoption of zoning directly in the hands of the county board of supervisors.

### Source of Zoning Powers

Zoning ordinances are adopted under the police powers delegated to the local governing body for the purpose of preserving and promoting the health, safety, morals, and general welfare of the public. These general objectives are achieved by establishing zoning regulations designed to "lessen congestion on the streets, to secure safety from fire, *flood,* panic, and other dangers; to promote health and the general welfare; to provide adequate light and air; to prevent the overcrowding of land; to avoid undue concentration of population; to facilitate the adequate provision of transportation, water, sewerage, schools, parks, and other public requirements," as stated in Section 358A.4 and Section 414.3, of the Iowa Code. In 1965 the enabling legislation was amended to make "safety from floods" a specifically mentioned objective of zoning.

Zoning ordinances, as usually drafted, consist of three basic parts.[2,3] These are the text, the schedules of district regulations, and the official zoning map. The "text" contains the administrative procedures for adopting, enforcing, and amending the ordinance. The "schedule of district regulations" contains the specific development standards and requirements for each of the zoning districts established by the ordinance, and the "official zoning map" delineates the boundaries of the districts. It is in these latter two sections that flood plain protection or regulation provisions are incorporated.

## Flood Plain Zoning Techniques

Several techniques are appropriate, but they vary in sophistication, depending upon data and administrative personnel available.[10] The simplest method is to incorporate a special condition regulation, applicable to all districts, into the supplementary district regulations of the ordinance. Such a clause might read: "The first floor of all structures shall be $x$ feet above the level of flood waters that may be expected to occur once in every $y$ years." The determination of whether a proposed building complies with the regulation is left to the judgment of the administrative officer, assisted perhaps by the city or county engineer. This requires that they have available either past flood records or a determination of potential flood levels for the affected areas. It does not require delineating on the district map a specific area subject to flooding. Frequently these provisions serve as stopgap measures until additional studies can be completed, which then permit more detailed methods to be introduced.

A more advanced technique is the delineation of a specific "flood plain district" on the official zoning map and the adoption of a separate set of district regulations. Such regulations usually limit land uses to open space, recreation, agriculture, and seasonal or nonpermanently occupied structures. More accurate topographic information is required to determine limits of the district accurately, but marginal areas perhaps can be left to the board of adjustment for final determination if and when the boundary may be questioned. In areas where industrial and commercial uses may already exist or are the logical uses of some areas subject to overflow, this single district does not suffice.

The "overlapping" or "blanket" zone technique can be used where flood plains are subject to several different categories of use. In this method the blanket zone covers that portion of the other districts in the ordinance which are located in the flood plain area, and specific requirements are established for various types of uses that may be in the flood plain in terms of location, elevation, occupancy, and others.

The "two-zone" concept carries the blanket-zone approach a step farther by creating a valley channel and valley plain area along the stream or watercourse. Ordinance provisions set floodflow elevations and encroachment limits for the valley channel, to permit an unobstructed flow of flood waters through the area and to establish the limits and elevations within which construction may be allowed on the valley plain. The ordinance might permit land in the valley plain to be filled prior to construction of buildings, but the ordinance seeks to keep the valley channel free from uses that would impede the flow of floodwaters.[5,7,14]

In the most refined of the several techniques the flood plain is separated into three zones: prohibitive, restrictive, and warning, as explained and defined by Kates and White:[8]

1. *Prohibitive*—that zone where any encroachment would, without clear justification to the contrary, be presumed to be against the public interest.
2. *Restrictive*—that zone where it would advance the general land-use and water-use aims of the community to restrict uses in relation to flood hazard.
3. *Warning*—that zone where it would be in the interest of property managers to receive warning of the risks involved but in which restriction is not deemed desirable.

In terms of flood plain areas, these zones pertain respectively to (1) the floodway; (2) the area landward of the floodway but which would be inundated by the design flood; and (3) the areas even more landward from the stream, which would now be inundated by floods greater in magnitude than the design flood.[1]

This last approach, a three-zone classification, places greater emphasis on the public interest in flood plain regulation, rather than on restrictions designed to protect the individual occupant. Both are justifiable objectives and should be the basis for preparing flood plain regulations regardless of the technique used.

### Types of Regulations

Within the district regulations of the ordinance there are a number of standards, restrictions, and guidelines that may be incorporated to achieve control of the flood plain, depending upon the detail desired.[10,14] As a minimum, however, the ordinance should spell out the type, location, and elevation of uses to be permitted. In addition, specific requirements might be established for the type, location, alignment, and materials to be used in fences and other structures; the construction, elevation, and location of levees, fills, and other protective measures; and the types, location, and quantity of materials permitted in open storage yards.

# BUILDING CODES, LOCAL ZONING, SUBDIVISION CONTROL 189

Planned-unit development (PUD) provisions are becoming increasingly popular and should include a site plan review and check of proposed grades, building elevation, and uses in the flood plain. Encouraging the eventual elimination of nonconforming uses, particularly those which may obstruct the floodway, by controlling reconstruction and requiring removal after a reasonable amortization period should also be considered.

Also, Section 358A.24 and Section 414.21 of the Iowa Code provide that:

> Wherever any regulation proposed or made under authority of this chapter relates to any structure, building, dam, obstruction, deposit or excavation in or on the flood plains of any river or stream, prior approval of the Iowa Natural Resurces Council shall be required to establish, amend, supplement, change or modify such regulation or to grant any variation or exception therefrom.

In county zoning, it should also be recognized that while farms and agricultural structures are generally excepted from zoning regulation they are subject to flood plain regulation included in such ordinances. Section 350A.2 was amended in 1965 and now reads:

> FARMS EXEMPT. No regulation or ordinance adopted under the provisions of this chapter shall be construed to apply to land, farm houses, farm barns, farm outbuildings or other buildings, structures, or erections which are primarily adapted, by reason of nature and area, for use for agricultural purposes while so used; *provided, however, that such regulations or ordinances which relate to any structure, building, dam, obstruction, deposit or excavation in or on the flood plains of any river or stream shall apply thereto.*

Finally, in preparing zoning ordinances and flood plain regulation, it is essential that (1) adequate data on topography, flood potential and existing land uses be obtained; (2) all regulations be reasonable, fair, and uniformly applied to similarly situated lands; and (3) the ordinance be in accordance with and designed to achieve the long-range objectives of a comprehensive plan.

## Legislation

Additional legislation is needed in Iowa to make zoning more effective and to protect areas not currently zoned. At the present time adoption of zoning must be by the city or county legislative body. Unfortunately, local governments have not readily assumed this responsibility, though towns have had 45 years and counties 12 years to do so. Consequently, extraterritorial zoning control of one, three, or five miles for cities and towns—as exercised in other states—and state zoning of rivers, lakes, state parks, state and federal highways, and the adjacent lands—one to five miles around them, perhaps—should be permitted. If local governments do not assume their responsibilities, then the next higher unit of government traditionally does it for them, with a corresponding loss of local control. More local communities in Iowa should participate in planning programs and enact local zoning ordinances and other regulatory measures.

## SUBDIVISION REGULATIONS

Subdivision regulation ordinances provide cities and towns with another tool to guide development in flood plain areas.[10] Under Chapter 409 of the Iowa Code, the municipal governing body must approve subdivision plats within its corporate limits, and if the community has an official city plan commission, within one mile of the corporate limits. A subdivision is defined in the act as the division of land into three or more parts.

A subdivision regulation ordinance establishes procedures for preparing, submitting, and processing subdivision plats; sets subdivision design standards; and requires certain improvements to be made as a condition of approving the plat. A plat may not be filed or lots sold from it until and unless it has been approved by the city council and city plan commission.

Review and approval procedures can protect against flood plain encroachment. These include checking plans to be certain that grading and storm drainage proposals are adequate to accommodate stormwater runoff and that building sites do not encroach upon the flood plain. Design standards should encourage the preservation of natural drainage courses to avoid costly storm sewer installation. Building lines and elevations can also be established as a plat restriction, where required to be certain that structures on individual lots will be above expected high-water elevations. Requiring restrictive covenants to include appropriate limitations or guidelines on filling, construction, and property maintenance, so that natural drainage patterns are protected, is an additional technique that needs to be considered.

Counties do not have subdivision regulation authority in Iowa, and enabling legislation to grant this is urgently needed. Another legislative change that would improve the ability of local government

# BUILDING CODES, LOCAL ZONING, SUBDIVISION CONTROL

to guide new developments effectively would be the redefinition of a subdivision as the "division of any parcel of land into two or more parts for other than agricultural purposes." The extension of extraterritorial subdivision controls to three or five miles for cities and towns would also be desirable.

## OTHER CODES AND ORDINANCES

In addition to zoning and subdivision regulation ordinances there are several other codes and ordinances that may be adopted which may contain provisions useful to promoting sound flood plain management. Among these are building codes, housing codes, plumbing codes, and mobile home park ordinances.

Building codes can be adopted in Iowa by all cities and towns under Section 368.9 of the Iowa Code and by counties over 25,000 population under Section 332.3 of the code. Usually a national model code, such as that of the Building Official's Conference of America,[4] International Conference of Building Officials,[6] or Pacific Coast Building Code Conference,[11] is adopted by reference and modified to fit local conditions.

Flood plain provisions can be added if not already contained in the basic code. Among the desirable provisions would be the establishment of floodproofing standards for structures in the flood plain and the requiring of construction of dikes to prevent hazards that could result from bulk petroleum, fertilizers, or other flammable or noxious materials stored in the flood plain.[10,12] Such clauses could reasonably be retroactive and require that all existing structures conform within a certain period of time after adoption.

A building code can also incorporate standards for structural materials, foundations, and anchorage of structures on the flood plain.[12] It also may dictate the placement and orientation of signs, buildings, and fences to minimize the impediment of floodflows.

Additional legislation is needed in Iowa to remove the population limit on counties, so that all counties which need building codes would have the authority to adopt them.

Housing codes can also incorporate desirable provisions to protect the occupants of residential structures from injury, loss of life, or property damage during floods. In addition to providing for sound construction and safe elevations, housing codes should ensure occupants of access to escape routes and emergency services and protect them from isolation or entrapment by flash floods.

Plumbing codes are useful in preventing indirect flooding from backup of sewers and in protecting water supplies and sanitary sewage disposal systems both for the individual and the community.

Mobile home park ordinances should provide for site plan, drainage, and lot elevation control. Submission of applications and receipt of approval should be required to ensure acceptance of a location safely above flood elevations and adequately drained in time of heavy stormwater runoff.

A final tool that can be used by local governments is that of acquiring floodway easements and/or development rights. Iowa needs clarifying or new permissive legislation in this area. By compensating owners for certain use or development rights, areas subject to flood could be maintained as open space yet left in the hands of private ownership for some compatible productive use. This would minimize the cost for the community, an all important consideration at this time when there are ever-increasing demands for the tax dollar.

## ADMINISTRATION

All the above codes and ordinances must be properly administered and enforced if they are to be effective. Most of them require some technical knowledge, and all require adequate records and communication with the public. For the small community which needs the protection afforded by these codes, this creates a difficult problem. To resolve this, a county-wide or regional approach to the administration of building and development codes has much to offer.

The intergovernmental cooperation act, a part of Chapter 83 of the Iowa Code, would permit the cities, towns, and county governments jointly to establish a regional code enforcement and inspection office and to contribute to its support. The result would be (1) uniform policies, standards, and procedures for the region; (2) a qualified full-time administrator; (3) adequate records; (4) an accessible source of information; and (5) good communication with the public at a cost every governing body could afford. A central office, with a mobile inspection and permit unit on the order of the traveling library or bookmobile, would enable the administrator and a small staff to cover the region on a regular rotating schedule. Hopefully, some communities will soon try this approach.

## CONCLUSIONS

The different types of local regulatory measures available today to our Iowa cities, towns, and counties have been outlined. The responsibility to adopt and enforce these measures rests squarely with the local legislative body. They can not, and the public should not let them, "pass the buck" or remain ineffective and unresponsive.

The additional enabling legislation needed at the state level has been discussed. It is believed that *all* this must be achieved if our communities are to keep up with the demands of these times.

Zoning ordinances, subdivision regulation ordinances, building codes, housing codes, plumbing codes, and acquisition of easement and development rights are the tools we have to protect the floodways from man, to protect man from floods, and to protect man from himself. We should make use of them *now*.

## REFERENCES

1. American Society of Civil Engineers. Guide for the development of flood plain regulations. Progress rept., Task Force on Flood Plain Regulations. *Proc. Am. Soc. Civil Eng., Hydraulics Div.* 88, no. HY5, Paper 3264 (Sept. 1962).
2. American Society of Planning Officials, Planning Advisory Service. *Zoning ordinance definitions.* Chicago. 1955.
3. Bair, F. H., and Martley, E. R. *The text of a model zoning ordinance, with commentary.* 2nd edit. Chicago: Am. Soc. of Planning Officials. 1960.
4. Building Officials Conference of America. *Basic building code.* New York. 1965.
5. Howe, J. W. Modern flood plain zoning ordinance adopted by Iowa City. *Civil Eng.* 33, no. 4 (Apr. 1963): 38–39.
6. International Conference of Building Officials. *Uniform building code.* Los Angeles. 1964.
7. Iowa City, City of. Ordinance 2238. 1962.
8. Kates, R. W., and White, G. F. Flood hazard evaluation. In G. F. White, ed., *Papers on flood problems.* Dept. of Geog., Res. Paper 70. Chicago: Univ. of Chicago Press. 1961.
9. Love, S. K. *Basis for urban developments, water.* Selected Papers from Am. Soc. of Planning Officials National Planning Conference, Seattle, Washington, 1963. (Chicago, 1963): pp. 47–56.
10. Murphy, F. C. *Regulating flood plain development.* Dept. of Geog., Res. Paper 56. Chicago: Univ. of Chicago Press. 1958.
11. Pacific Coast Building Officials' Conference. *Uniform building code, 1960.* Los Angeles. 1960.
12. Sheaffer, J. R. *Introduction to flood proofing, an outline of principles and methods.* Chicago: Univ. of Chicago Press. 1967.
13. U.S. House of Representatives, Committee on Public Works. *A unified national program for managing flood losses.* Rept. of the Task Force on Federal Flood Control Policy. House Document 465, 89th Cong., 2nd sess. 1966.
14. Wisconsin Department of Natural Resources, Division of Resource Development. *Model flood plain zoning ordinance for a city or village.* Madison. 1967.

East Nishnabotna River at Atlantic, 1958. Courtesy Phil Chinitz, Atlantic **News-Telegraph.**

**Part 6**

*Flood plain information studies are intended to furnish advice and to encourage those affected to help themselves.*

Corps of Engineers

# AGENCY FUNCTIONS AND SERVICES FOR FLOOD PLAIN MANAGEMENT ASSISTANCE

# 17

# FLOOD PLAIN MAPPING BY THE
# U.S. GEOLOGICAL SURVEY

### D. W. Ellis

THE U.S. GEOLOGICAL SURVEY (USGS) was engaged in water resources investigations as early as 1890. For more than 50 years the mutual interest of the state and federal governments in water-related problems have been implemented in the USGS by cooperative investigations. The annual appropriation bills concerning the USGS provide, "that no part of this appropriation shall be used to pay more than one-half the cost of any topographic mapping or water resources investigations carried on in cooperation with any state or municipality." More than 300 state and local agencies participate in the nationwide water resources investigations through financial cooperation with the USGS. The current trend is for increased cooperative programs, as the nation is becoming more aware of the need to find solutions to the many water problems which have arisen in the United States. Insofar as surface water problems are concerned, these include the recent water shortage in the northeastern states, widespread stream

D. W. ELLIS is Hydrologist, U.S. Geological Survey, and serves as Chief of the Special Studies Section in Champaign, Illinois. Publication of this paper authorized by the Director, U.S. Geological Survey.

pollution, periodic flooding, and flood plain management.

For many years the water problems of the country were such that the primary emphasis in the survey's program was the collection and publication of basic streamflow data in annual water-supply papers (WSP).[1] More recently, the need for solutions to increasingly complex water problems has demanded greater attention to analytical, interpretative, and research phases of water resources investigations, as well as to the form in which the data should be published, to enhance their usefulness in solving particular problems.

## FLOOD PLAIN OCCUPANCY

One problem receiving much attention is caused by man's occupancy of flood plains. Failure to recognize that their natural function is to carry away excess water during floods has led to haphazard development on flood plains resulting in an increase in hazards. The average annual flood damage in the United States has tripled since the beginning of the century, even though billions of dollars have been spent for flood protection works.[2] It is reported that for every $6 spent by the federal government for flood control, $5 is spent by the public expanding onto flood plains. These figures serve to point out that constructing flood control works is not a complete solution to flood problems.

In recent years much thought has been given to flood plain planning and regulation as being more effective preventive measures for reducing losses. Regulation, including zoning, would reduce damage by requiring that the flood plain be used for purposes not subject to severe flood damage or by requiring protective measures, landfill, or other controls. Effective flood plain management requires a knowledge of the magnitude and frequency of floods to be expected and of the areas that will be inundated. The series of flood inundation reports (hydrologic atlases) of the USGS are intended to aid individuals and agencies responsible for making planning decisions by providing this needed hydrologic data.

## FLOOD STUDIES AND REPORTS

The USGS has published descriptions of several hundred flood events in water-supply papers. Beginning in 1950, annual water-supply papers have been published that summarize significant floods throughout the year. Separate papers have been prepared for outstanding events, such as the 1951 floods in Kansas,[3] the 1953 floods in northwestern Iowa,[4] the 1954 floods in northern and central Iowa,[5] the 1955 floods in the northeastern states, and the 1955–1956 floods in the far western states.[6] More recently, a report was prepared for the Mississippi River flood of March–May, 1965, and is in process of publication as USGS WSP 1850-A.[7]

The first USGS flood report (WSP 88), which describes a flood on the Passaic River in New Jersey, includes a topographic map of an area inundated by the floods of February–March, 1902. Several other flood reports published as water-supply papers have also included maps and sketches showing inundated areas for specific floods.[8] Thus, flood mapping is not new to the USGS. However, just a map showing inundated areas is not sufficient for making sound decisions on flood plain development. What is the chance of a flood reaching a particular location or what is the risk involved by permitting the construction of a building at a particular location? This is the type of question that planners need to consider.[9]

## FLOOD MAPPING PROGRAM

In its continuing effort to adapt its reports to changing needs, the USGS undertook a pilot project to develop a report on floods that would be simple and concise, yet factual and informative. The report needed to be presented in language that could be understood by people with nontechnical backgrounds and could be published in a form suitable for widespread distribution. From this undertaking came the report, "Floods at Topeka, Kansas," which was published in 1959 as Hydrologic Investigations Atlas 14.[10] The report is contained on a single sheet of paper about 32 x 30 inches, which is a convenient size for making field investigations. Components of the report are the topographic map, a history of past floods, a flood frequency diagram, flood profiles, an oblique aerial photograph taken during the flood, and a brief explanatory text. These features were adopted for general use in subsequent flood mapping studies in urban areas, each published as a hydrologic investigations atlas.

### Details of the Topeka Atlas

A standard USGS 7.5-minute topographic quadrangle provides the base map (scale: 1 inch = 2,000 feet). A blue overlay is printed to show the extent of inundation by the Kansas River during the

1951 flood. The high-water profile corresponding to the flooded area is shown, with profiles of several lower floods. The frequency curve indicates the average interval of time that should be expected to elapse between floods exceeding a given elevation at the Topeka Avenue gaging station. The flood histogram not only provides a history of past floods but also emphasizes the fact (frequently misunderstood) that floods do not occur regularly according to their recurrence intervals. For example, the frequency curve shows that a stage of 25 feet corresponds to a recurrence interval of about 7 years, whereas the histogram shows that the 25-foot stage was not exceeded during the 26-year period 1909–1934, but that it was exceeded 4 years in succession during 1941–1944. A convenient and perhaps a more easily understood way of referring to frequencies of floods is in terms of their probabilities. The probability of a flood occurrence is virtually the reciprocal of its recurrence interval, particularly for recurrence intervals greater than about 10 years. For example, there is about a 4 percent chance that a 25-year flood would occur in any one year and likewise about a 2 percent chance that a 50-year flood would occur in any one year.[11]

Although the frequency curve is applicable for a specific location, by use of the profiles the probable frequency can be transposed to other points along the stream with reasonably reliable results. As flood profiles tend to be parallel, a profile of any given frequency can be drawn parallel to and above or below an existing profile by the amount indicated by the stage-frequency diagram.

In addition to showing the areas inundated by the maximum known flood, the report also provides sufficient information for delineating zones of various risks of flooding. Some types of development might be restricted from the flood plains on the basis of the maximum known floods or perhaps limited to some higher elevation, whereas other types of development might be permitted in the flood plains but be restricted by a risk concept. This kind of decision is made by those agencies or units of government which are vested with the authority and responsibility of formulating policies for effective flood plain regulation. The flood atlases of the USGS are intended to provide the necessary hydrologic data and to serve as a tool for those charged with planning responsibilities.

### Additional Flood Mapping Investigations

The Topeka report was favorably received in other parts of the United States, and the magnitude of the survey's flood mapping program is accelerating in response to widespread needs and requests. Excluding the northeastern Illinois program, which is discussed herein, 53 similar reports either are published or are in preparation in 16 states and Puerto Rico (12 in Ohio, 8 in Puerto Rico, 5 in New Jersey, 3 in Texas, 2 each in 4 states and 1 each in 9 states). One such report in Iowa was prepared and published in 1963 for "Floods on Des Moines River, Raccoon River, Walnut Creek, and Four-Mile Creek at Des Moines, Iowa, in 1947, 1954, and 1960."[12]

Soon after the Topeka report was distributed, officials of the Northeastern Illinois Planning Commission made inquiries about the possibility of negotiating a flood mapping program for the Chicago metropolitan area. The USGS prepared a report for the area that would serve as a prototype for the possible cooperative program. This report was published in 1961 as HA-39, "Floods near Chicago Heights, Ill." The elements of this report are similar to those for the Topeka report: a map showing the areas inundated by past floods, corresponding flood profiles, flood frequency curves, a histogram, and a concise explanatory test. A distinct difference in the two reports is noted: Whereas the Topeka report described the flooding only along the main stream, the Chicago Heights report described flooding along all streams on the map.

The staff of the Northeastern Illinois Planning Commission recognized the potential of this type of report as a planning tool. If planning agencies and municipalities within the region used such reports for establishing building and zoning regulations, the creation of many flood problems could be avoided by limiting the use of flood plains.

Discussions with the planning commission resulted in an extensive five-year cooperative program whereby flood atlases would be prepared for 43 7.5-minute quadrangles. This initial program is the largest yet undertaken by the USGS. During the third year of the program, the commission requested the survey to prepare flood atlases for 19 additional quadrangles during a three-year extension of the program. Flood maps would then be available for essentially the entire metropolitan area, covering approximately 3,600 square miles. The initial 43 atlases were completed on schedule in June, 1966, and since that time 10 atlases have been completed under the extended program.[13]

The arrangements for the local share of finances for this program are unique. The planning commission did not have funds that could be allocated to the program, but it arranged contractual agreements with each of the six counties involved, whereby each county contributed a proportionate share of cooperative funds.

### Flood Problems Along Small Tributaries

It was desirable in the northeastern Illinois program to flood-map the seemingly insignificant tributaries, as much of the development taking place in the metropolitan area was along these streams. The initial effort of the program considered most important was the establishment and operation of a network of crest-stage gages to record high-water elevations to aid in developing flood profiles. About 450 gages are presently maintained in northeastern Illinois, of which 394 were established as part of the flood mapping program. The gages define lower profiles and, in the event of a major flood that exceeds those previously mapped, much of the data required for updating the maps would already be available.

The program was designed to map the highest floods that could be defined rather than to compute hypothetical floods of selected frequencies. The planning commission foresaw the possibility that zoning regulations based on the maps would receive opposition from those adversely affected. It was believed that historical data would be more acceptable to the courts than hypothetical computations. The Highland Park atlas (HA-69) was involved in a condemnation suit almost immediately after its release. A park board undertook to purchase a portion of the flood plain to develop as a park and offered to buy at the going rate for flood plain property. This land was controlled by a real estate speculator, who claimed the property was suitable for higher uses and demanded a correspondingly higher price. When the park board introduced the flood map as evidence, the realtor objected on the grounds that it was hearsay evidence—the man who made the map did not personally witness the flood. The court was about to uphold the realtor's position until the park board produced a witness who had personally observed the flood and testified as to the validity of the map. The map was then admitted as evidence. It was apparent that this particular court would have rejected a map based on hypothetical computations.

In addition to showing the inundated areas, all stream-gaging stations are plotted on the flood maps also. The subwatershed drainage boundaries are shown for each station, and the sizes of the drainage areas are given in the text of the report. This facilitates the determination of the drainage area at other points along the streams where it might be needed for planning and design of highway crossings, flood control structures, channel improvements, and the like.

## DISTRIBUTION AND USE OF FLOOD ATLASES

An important element of any program is making use of the results. The Northeastern Illinois Planning Commission has done an excellent job of publicizing the program and educating local planners, municipal and county officials, and the public in general as to how the reports can be put to use. In addition to publicity in the newspapers, the commission issues a number of publications in which the flood mapping program is emphasized. Examples of such publicity are (1) flyers entitled "Recent Planning Publications" which are issued periodically, (2) a monthly newsletter, and (3) annual reports.

In the early years of the program when a report became available, the commission arranged a public meeting in the immediate vicinity of the area mapped. The meeting was given advance publicity in local newspapers. State, county, and municipal officials and local planners received special invitations to attend. At the meeting, contents of the report were explained by personnel of the USGS, and planning commission personnel outlined some specific ways for applying the data to solve existing flood problems and to minimize the creation of new ones. Several of these meetings were held in each county.

The planning commission serves as the local distributor of the published atlases. It has a standing order for 200 copies of each, in addition to the 200 it receives automatically by being the cooperating agency. Widespread interest in the atlases is evidenced by the fact that for several the initial supply of 400 copies was soon exhausted. Some factors that may account for the interest in the reports are:

1. The publicity that the program receives.
2. Increasing awareness of flood problems by the public.
3. The omission of technical jargon from the language of the reports, which makes them understandable and useful to the average citizen without technical training.
4. The minimum cost of only $.75 per copy.

Merely providing information as to where floodwaters have reached and how often such floods may be expected does not necessarily mean that everyone will avoid construction in flood plains. Some people may not be easily convinced of flood risks unless they have actually seen a flood occur. Many subdivisions and other types of developments have been constructed in flood plains and sold to unwary citizens during a flood-free period. The information, however, can be effectively used through proper land-use planning or zoning.

Examples of the use of flood atlases in northeastern Illinois can be noted. Several municipalities have established zoning regulations for residential buildings on the basis of the flood atlases. The county forest preserves and park districts used the atlases in purchasing land subject to flooding, thus eliminating possible encroachments on the flood plains. The metropolitan sanitary district of greater Chicago, which has drainage and sewerage responsibilities in Cook County, has a regulation prohibiting sanitary sewer connections to new buildings in the flood plains described by the atlases, thereby discouraging further developments there. Many mortgage companies use the atlases to determine the flood potential of property for which loan applications have been received. Appraisers for the Veterans Administration and the Federal Housing Administration must refer to an atlas, if available, for the property on which a loan guarantee has been requested.

## SUMMARY

One way to indicate the economic benefits of a project is to refer to its benefit-cost ratio.[14] In a paper presented to the Highway Research Board in 1964, John Sheaffer estimated that the flood mapping program in Illinois has returned benefits that exceed the cost by a ratio of 40 to 1.[15]

The flood atlases of the USGS present hydrologic data concerning the extent, depth, and frequency of flooding that are essential for an appraisal of hazards involved in development of flood plains. The atlases are useful in establishing programs of flood plain management, preparing building and zoning regulations, locating waste disposal facilities, purchasing open space, developing recreational areas, and managing surface water in relation to groundwater resources.

## REFERENCES

1. Prior to 1961, annual water-year stream flow data were published by the USGS in several parts in reports known as *Surface water supply of the United States* (Washington: USGPO). These surface water records

were summarized in 1950 and again in 1960 and include for Iowa: (a) Compilation of records of surface waters of the United States through September 1950. *Part 5. Hudson Bay and Upper Mississippi River basins,* USGS WSP 1308, 1958; *Part 6-A. Missouri River basin above Sioux City, Iowa,* USGS WSP 1309, 1958; and *Part 6-B. Missouri River basin below Sioux City, Iowa,* USGS WSP 1310, 1958. (b) Compilation of records of surface waters of the United States, October 1950, to September 1960. *Part 5. Hudson Bay and Upper Mississippi River basins,* USGS WSP 1728, 1964; *Part 6-A. Missouri River basin above Sioux City, Iowa,* USGS WSP 1729, 1964; and *Part 6-B. Missouri River basin below Sioux City, Iowa,* USGS WSP 1730, 1964. Beginning in 1961, streamflow records and related data were published in annual reports on a state-boundary basis. For Iowa these are currently being published annually as *Water resources data for Iowa: Part 1, surface water records* and are distributed through the Water Resources Division, USGS, Iowa City, Iowa. Five-year summaries will be made on the former report basis. Flood discharge records for Iowa are also summarized in H. H. Schwob, *Magnitude and frequency of Iowa floods,* Bull. 28, pt. 2 (Ames: Iowa Highway Res. Board, 1966).

2. A summary of the nation's problems in controlling flood losses is found in U.S. House of Representatives, Committee of Public Works, *A unified national program for managing flood losses,* Rept. of the Task Force on Federal Control Policy, House Document 465, 89th Cong., 2nd sess., 1966.
3. See Water Resources Division, USGS, *Kansas-Missouri floods of July, 1951,* USGS WSP 1139 (Washington: USGPO, 1952).
4. See J. V. B. Wells, *Floods of June, 1953, in northwestern Iowa,* USGS WSP 1320-A (Washington: USGPO, 1955).
5. See I. D. Yost, *Floods of June, 1954, in Iowa,* USGS WSP 1370-A (Washington: USGPO, 1958).
6. See W. Hofmann and S. E. Rantz, *Floods of December, 1955–January, 1956, in the far western states,* USGS WSP 1650-B, pt. 2 (Washington: USGPO, 1963).
7. See also for Iowa areas H. H. Schwob and R. E. Myers, *The 1965 Mississippi River flood in Iowa,* open-file report (Iowa City: USGS, 1965).
8. For instance, inundated areas at Sioux City during the 1953 Floyd River flood are shown in Wells, *Floods of June, 1953,* p. 8.
9. A comprehensive study by the USGS which showed how hydraulic and hydrologic data can be used in evaluating the flood potential and the associated risk of flood plain occupancy was reported by S. W. Wiitala, K. R. Jetter, and A. J. Sommerville, *Hydraulic and hydrologic aspects of flood plain planning,* USGS WSP 1526 (Washington: USGPO, 1961).
10. See U.S. Geological Survey, *Floods at Topeka, Kansas,* Hydrologic Investigations Atlas HA-14, Washington, 1959.
11. For a summary of flood frequency analytical methods used by the USGS, see T. Dalrymple, *Flood frequency analyses,* USGS WSP 1543-A (Washington: USGPO, 1960). Also M. A. Benson, *Evaluation of methods for evaluating the occurrence of floods,* USGS WSP 1580-A (Washington: USGPO, 1962); and M. A. Benson, *Factors influencing the occurrence of floods in a humid region of diverse terrain,* USGS WSP 1580-B (Washington: USGPO, 1962). Floods in Iowa have been studied and reported by Schwob, *Magnitude and frequency of Iowa floods.*
12. See R. E. Myers, *Floods at Des Moines, Iowa,* Hydrologic Investigations Atlas HA-53 (Washington: USGS, 1963).
13. For a typical report on small tributary areas, see D. W. Ellis, H. E. Allen,

and A. W. Noehre, *Floods of Elmhurst quadrangle, Illinois,* Hydrologic Investigations Atlas HA-68 (Washington: USGS, 1963).
14. The concepts of benefits and costs and the use of the benefit-cost ratio by the federal agencies are summarized in: U.S. Senate, *Policies, standards, and procedures in the formulation, evaluation, and review of plans for use and development of water and related land resources,* Document 97, 87th Cong., 2nd sess., 1962. Additional concepts on economic use of flood plains are included in House Document 465, 1966 (see note 2, supra).
15. See J. R. Sheaffer, *The use of flood maps in northeastern Illinois,* Highway Res. Board 58 (Washington: Nat. Acad. Sci., 1964): 44–46. For additional information concerning the advantages of the flood mapping program, see T. Dalrymple, *Flood mapping program of the U.S. Geological Survey,* Highway Res. Board 58 (Washington: Nat. Acad. Sci., 1964): 33–41.

# 18

# ASSISTANCE THROUGH THE FLOOD PLAIN MANAGEMENT SERVICES OF THE CORPS OF ENGINEERS

### John N. Stephenson

THE PURPOSE of this chapter is to outline and discuss the technical assistance which the U.S. Army Corps of Engineers can provide through its flood plain management services (FPMS) program. This is a comprehensive program; and with the cooperation of all concerned federal, state, and local governmental units it can assist in reducing the staggering flood loss that our nation suffers each year. The federal interest in this matter is beyond doubt, and efforts to cope with the problem will be unsparing. A very great responsibility for success of the program rests upon state and local governments and upon individual property owners in flood hazard areas.

## NEW NATIONAL APPROACH

The development of the FPMS program of the Corps of Engineers began in 1960, upon the direction of Congress. The need for new

JOHN N. STEPHENSON is Assistant Chief, Planning Division for Flood Plain Management Services, U.S. Army Engineer Division, North Central Division, Chicago, Illinois.

approaches in the problem of flood plain use and development was first brought to the attention of the Congress by the U.S. Senate through the Select Committee on Water Resources in 1959 and 1960. Three publications were specifically concerned with floods, flood problems, and the need for flood plain management.[1]

Congressional action in 1960 established and initiated a flood plain information program. Section 206 of the Flood Control Act, approved July 14, 1960, Public Law 86–645, as amended, authorizes the Secretary of the Army, through the Chief of Engineers, to compile and disseminate information on floods and flood damage potentials and general criteria for guidance of federal and nonfederal interests and agencies in the use of flood plain areas. The Corps of Engineers carries out its FPMS program under this basic authorization.[2] This program was greatly expanded in August, 1966, by a task force report, printed as House Document No. 465, and the issuance of Executive Order 11296.

### House Document No. 465

A task force on federal flood control policy was appointed by the Federal Bureau of the Budget late in 1964 to study the flood situation in the United States and to recommend action to alleviate flood losses affecting the national economy. The task force report, titled "A Unified National Program for Managing Flood Losses," was transmitted to Congress by President Johnson in August, 1966.[3] In his letter of transmittal President Johnson strongly endorsed the report, which calls for different emphasis by the federal, state, and local agencies and by property owners. This would be achieved through the combined actions of all federal agencies dealing with floods and use of flood plains. States would play a pivotal role in coordination and in the flow of information.

### Executive Order 11296

At the same time that President Johnson transmitted the task force report to Congress, he also issued Executive Order 11296, which became fully effective January 1, 1967.[4] In accordance with this order, all federal agencies are now evaluating flood hazards in locating federally owned or financed buildings, roads, and other facilities; in disposing of federal lands and properties; and in the administration of federal grant, loan, or mortgage insurance programs involving construction of buildings, structures, roads, or other facilities. The federal government is thus taking the lead in precluding the uneconomic, hazardous, or unnecessary use of flood plains. The agencies are also required to floodproof existing federally owned facilities whenever practical and economically feasible. The order directs the Secretary of the Army, and the Tennessee Valley Authority (TVA) for those lands within the Tennessee River basin, to provide flood hazard information and guidance on floodproofing.

The federal agencies are cooperatively developing guidelines and criteria for evaluation of flood hazards, permitting uniform and equitable treatment of flood hazard problems throughout the nation. Some have issued interim instructions or regulations which will later be finalized, and the others are expected to issue such regulations when coordinated guidelines are available.

## CORPS OF ENGINEERS' FPMS PROGRAM

The expanded Corps of Engineers' FPMS program provides a means of obtaining the technical information needed by flood plain managers. The program operates in each of 47 division and district offices with nationwide coverage. The purpose is to make available to federal, state, and local governmental agencies the information, guidance, and advice on the flood hazard, permitting them to proceed with such planning, engineering studies, construction, and other action as may be necessary for wise use of flood plains.[5]

The program includes:

1. Preparation of flood plain information reports.
2. Provision of technical services and guidance.
3. Conduct of related research.
4. Comprehensive flood damage prevention planning.

Each of these items will be discussed in turn.

### Flood Plain Information Reports

Flood plain information reports are prepared upon request of state and local agencies to delineate flood problems in communities throughout the country. The Corps of Engineers has gathered much flood plain information during the conduct of its previously authorized responsibilities for flood control. Such information is assembled and organized under this program. When necessary, additional physical surveys and hydrologic studies are undertaken. A typical report includes maps or mosaics, profiles, charts, tables, photographs, and a narrative describing the extent, depth, and duration of flooding experienced in the past and expected in the future.

It is the responsibility of the state and local governmental agencies to publicize the information and put it to use through planning groups, zoning boards, private citizens, engineering and planning firms, real estate and industrial developers, and others to whom it would be useful.

Nearly every community has a flood problem. There are more than 5,500 communities with 2,500 population or more, and several thousand places with smaller populations.[6] This means there are probably in excess of 7,500 communities with flood problems. The Corps of Engineers, TVA, and other federal agencies have covered about 300 of these in flood plain information reports. The corps plans to prepare reports for about 200 additional communities this fiscal year and 250 or more annually thereafter.

### Technical Services

Technical assistance is given states and local governments in the preparation of flood plain regulations. Interpretation of flood data in the flood plain information reports, provision of additional data, suggestions for floodway areas and evaluation of their effect on flood heights, and related assistance are given planners and officials as they prepare and adopt flood plain regulations.

Technical assistance is given to other federal agencies, states, and local governments in evaluating and using flood data in making individual decisions concerning flood hazards. This includes brief, preliminary type flood plain information reports where necessary for specific sites. Necessary flood information and guidance is provided on request to permit wise decisions concerning locations of public buildings, subdivisions, and other land uses.

Technical assistance and guidance is also provided to other federal agencies, states, and local governments on floodproofing through the use of guides and pamphlets. Technical assistance, a very important part of the program, will be discussed in more detail later.

### Guides, Pamphlets, and Related Research

The research activity includes studies to improve methods and procedures for flood damage prevention and abatement and for the preparation of guides and pamphlets pertaining to floodproofing, flood plain regulations, flood plain occupancy, economics of flood plain regulations, and other related approaches to flood damage prevention. These activities are being coordinated with related programs of other federal and state agencies. The guides and pamphlets are made available for use by state and local governments, private citizens, and federal agencies in planning and in taking action to reduce the flood damage potential.

A pamphlet entitled *Introduction to Flood Proofing—An Outline of Principles and Methods* has been prepared under the joint sponsorship of the Corps of Engineers and the TVA.[7] Others will be issued from time to time. Another pamphlet entitled *Guidelines for Reducing Flood Damages* has also been completed.[8] Each of these is being given nationwide distribution.

The Corps of Engineers has received requests to develop floodproofing standards or specifications for use in flood plain management programs. Replies state that publications containing a greater variety of suggestions than presented in the cited pamphlet and examples of floodproofing techniques for various conditions and situations are planned. However, the extent to which this work will produce information that can be used in writing definitive standards or specifications is not known because of the wide differences in problems. It is encouraging, however, to note the interest that the floodproofing pamphlet has stimulated, as evidenced by the requests for additional information.

As an interim measure, consideration is being given to the preparation of guidance on inclusion of requirements for floodproofing in building codes and other regulations for use of land and construction of properties in flood plain areas. This can probably be completed in three to four months after the work is started. Preparation of the other pamphlets will require somewhat greater effort and time.

## Comprehensive Planning for Flood Damage Prevention

Comprehensive planning for flood damage prevention at all appropriate governmental levels is the composite and ultimate objective of the FPMS program. It is not a separable activity, but takes place through preparation and dissemination of flood plain information reports and through provision of technical advice and other forms of assistance to persons and governmental entities dealing with problems of flood hazard.

The purpose of these activities is to encourage and guide the best and safest possible use of flood plain lands for the benefit of the national economy and welfare. The program provides a highly desirable and useful tool in rounding out the national effort in treating flood problems and provides a direct complement to ongoing Corps of Engineers' flood control efforts using engineering works.

## THE CORPS OF ENGINEERS' FPMS PROGRAM IN IOWA

The general overall picture of the Corps of Engineers' FPMS program has been presented. The detailed manner in which it operates in the state of Iowa will be discussed below.

### Coordinating Agency

The coordinating agency in Iowa for this program is the Iowa Natural Resources Council (INRC). Requests for flood plain information reports and technical assistance should be submitted to that agency, which will in turn submit the request to the proper Corps of Engineers district office. The coordinating agency also assigns the priority of the flood plain information reports, requests for which the district office will forward to the proper division office for approval action. If the request is for technical assistance, the district office may handle it directly if the desired information is readily available or if the cost of obtaining it is not too great. Matters requiring approval of higher authority are forwarded to the office of the Chief of Engineers by the division office. In brief, the channel of communication for requests is (1) community, (2) state coordinating agency, (3) district office, (4) division office, and (5) office of the Chief of Engineers. Replies to requests are channeled back in the reverse order.

### Corps of Engineers' FPMS Organization in Iowa

The overall conduct and direction of the FPMS program in the Corps of Engineers are by a FPMS branch in the office of the Chief of Engineers in Washington, D.C., and by a FPMS unit in each division and district office.

The boundaries of two divisions and four districts lie within the state of Iowa, namely, the north central division with two districts, St. Paul and Rock Island and the Missouri River division with two

## FLOOD PLAIN MANAGEMENT SERVICES, CORPS OF ENGINEERS

districts, Omaha and Kansas City. The boundary of the north central division embraces those rivers and streams which flow into the Mississippi River, and the Missouri River division includes the drainage which enters the Missouri River.

As of 1968 Brigadier General Robert M. Tarbox was the division engineer of the north central division with an office in Chicago, Illinois. Brigadier General C. Craig Cannon was the division engineer of the Missouri River division, Omaha, Nebraska. Harold Bates was chief of the FPMS program for the Missouri River division. District personnel were:

| District | District Engineer | Chief, FPMS Program |
|---|---|---|
| Omaha district | Col. Wm. H. McKenzie III | Norman Gau |
| Kansas City district | Col. W. G. Kratz | Glenn Carriker |
| St. Paul district | Col. Richard J. Hesse | Harold Toy |
| Rock Island district | Col. Walter C. Gelini | Donald Davis |

Figure 18.1 shows the basin boundaries and the Corps of Engineers division and district boundaries.

### Flood Plain Information Reports in Iowa

The corps has completed five flood plain information reports in Iowa:

Indian Creek and Dry Creek, Linn County ....... December, 1964
Duck Creek, Scott County ......................... July, 1965
Prairie Creek, Linn County ..................... February, 1966
Skunk River and Squaw Creek, Story County ......... June, 1966
Cedar River at Cedar Rapids ................. December, 1967

The locations at which flood plain information reports have been completed in Iowa are shown in Figure 18.2.

As of 1968, one report for Bedford on the East Fork of the Hundred and Two River is being prepared by the Kansas City district. The Corps of Engineers has received only one application for further studies; therefore, the tentative fiscal year 1969 schedule includes only one study for Iowa. In areas where additional studies are needed, applications should be submitted by local agencies to the INRC. Upon receipt, these applications will be placed in the current fiscal year schedule by a priority classification, and studies will be made if or as soon as funds are available.

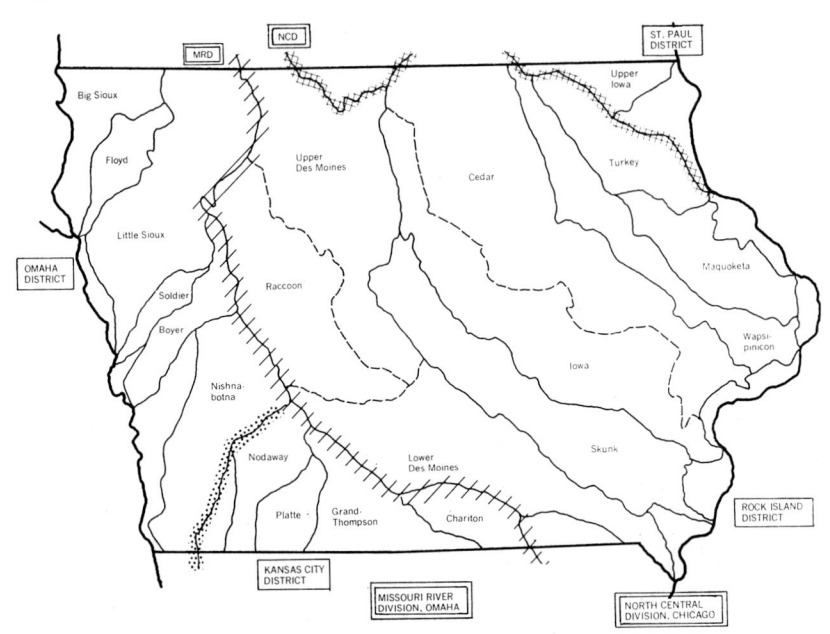

Fig. 18.1. Corps of Engineers district boundaries and headquarters for river basins in Iowa.

## Provision of Technical Service and Guidance

The Corps of Engineers' personnel have worked very closely with state and local agencies in the preparation of flood plain information reports. However, working relations could be considerably broadened from a statewide viewpoint by aid to local groups, upon their request, through the provision of technical assistance and guidance. Local requests for technical assistance provide an opportunity to become better acquainted with community problems and planning and are a means of increasing contacts with all levels of local governments. Such contacts provide a means of affording such guidance not only on flood plain matters but on all related local planning problems.

As mentioned earlier, the Corps of Engineers has completed five flood plain information reports in Iowa; one is now under preparation, and one application for a study is being processed. Based on these figures, one could estimate that within the next year seven areas will be covered by flood plain information reports. Review of the list of urban areas in Iowa with 2,500 or more inhabitants indicates there are approximately 70 that have some type of flood problem. The list also indicates that a number of these areas have been studied by the

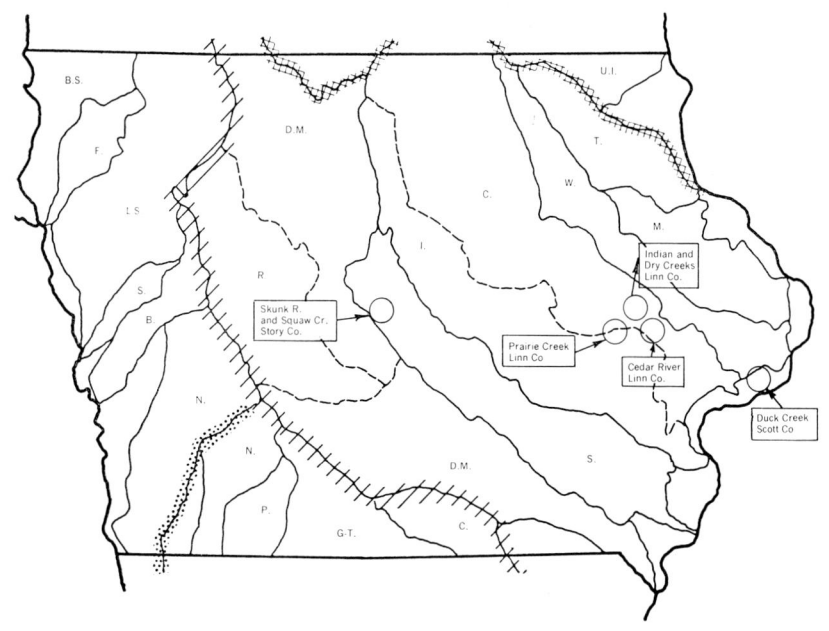

# FLOOD PLAIN MANAGEMENT SERVICES, CORPS OF ENGINEERS 215

Fig. 18.2. Completed flood plain information reports for communities in Iowa.

Corps in the interest of flood protection. A total of 23 reports found engineering works for flood protection to be economically unfeasible. Eight are now under restudy. In view of this information, it would appear that the information already obtained by the Corps could be used to advantage in attempting to define the flood hazard in these areas. This does not mean that all places would require a flood plain information report, but that with available information the Corps could provide valuable technical assistance. In this regard, it should be pointed out that the Corps will handle all requests for technical assistance and guidance as expeditiously as possible within the limitation of manpower and funds.

## Comprehensive Flood Damage Prevention Planning in Iowa

Comprehensive flood damage prevention is not a separable activity. It involves consideration of all measures available for prevention of flood damages, both structural and nonstructural, used separately or in combination. Structural measures would include levees, channel rectification, flood control reservoirs, floodproofing, and the like; while nonstructural measures would include zoning, subdivision regulations, flood forecasting, and so forth. Seldom is one independent measure the complete answer to the flood plain problem. An analysis of many alternative solutions will define the combination of measures that will provide the best solution.

Iowa has examples of most of the individual ways of preventing flood damage: reservoirs, such as Coralville; channel improvements, such as at Van Meter; levee systems along the Mississippi and tributaries; construction of elevated buildings, such as on the campus of Iowa State University; evacuation of the flood plain, which was considered at Littleport; the flood plain zoning at Iowa City; and others.

An example of a combination of structural measures would be the Saylorville Reservoir and the Des Moines levee system. An example of a combination of a structural measure and a nonstructural measure would be the Coralville Reservoir and flood plain zoning at Iowa City.

Flood control survey reports prepared by the Corps of Engineers have always considered alternative solutions to the flood control problem, including evacuation of the flood plain, which was considered at Littleport on the Volga River. However, in future survey reports nonstructural measures will be presented in more detail, particularly for those communities where structural solutions result in unfavorable benefit-cost ratios.

Finally, public awareness of the flood problem is an essential element of comprehensive flood damage prevention planning. Publicity regarding flood plain information reports and/or other technical studies should be directed toward planning groups, zoning boards, engineering and planning firms, and real estate and industrial developers as well as private citizens. Local bankers (home loan or industrial loan companies) should be advised as to the flood hazards on both developed and undeveloped flood plain areas within their community or regional area.

## SUMMARY

President Johnson's statements and directives and action by the Bureau of the Budget reflect the acceptance by the executive branch of the federal government of the principle of flood plain management. Action by the Congress in broadening the authorization and increasing the monetary limitation for the Corps of Engineers' FPMS program and in continuing the TVA local flood relations program reflects the mood and acceptance of the federal legislative branch.

States have, for years, recognized in varying degrees the need for a nationally coordinated program, with the states playing a greater role. Studies, papers, and seminars of the Council of State Governments since 1958 portray the nationwide thinking of state officials. The increasing number of states considering state legislation and statewide programs is further evidence of greater awareness and acceptance at that level.

The increasing number of cities and counties from the east coast to the west coast and from the Canadian border to the Rio Grande that are adopting flood plain regulations is proof that the approach is being accepted at the local level.

Each level of government has a role in proper flood plain management. The federal interest is unquestioned. But state and local governments and individual owners of properties in flood hazard areas also must meet their great responsibilities if there is to be improved management. All must work closely and diligently, so that the combined local-state-federal effort will bring about the wise use of water resources and flood plains.

The Corps of Engineers will continue to do its part in making the program a success. Its personnel are willing and anxious to assist in efforts to promote the wise use of flood plain areas.

## REFERENCES

1. See U.S. Senate, Select Committee on National Water Resources (Washington: USGPO): *Floods and flood control,* Committee Print 15, July 1960; *Flood problems and management in the Tennessee River basin,* Committee Print 16, Dec. 1959; and *River forecasting and hydrometeorological analysis,* Committee Print 25, Nov. 1959.
2. Details of the initial program are found in U.S. Army Corps of Engineers, *Flood plain information studies, water resource policies and authorities,* Manual EM 1165-2-111 (Washington: Chief of Engineers, 1961).
3. See U.S. House of Representatives, Committee on Public Works, *A unified national program for managing flood losses,* Rept. of the Task Force on Federal Flood Control Policy, House Document 465, 89th Cong., 2nd sess., 1966.
4. See Office of the President of the United States, *Evaluation of flood hazard in locating federally owned or financed buildings, roads, and other facilities, and in disposing of federal lands and properties,* Exec. Order 11296, 1966.
5. Land use and flood problems in flood plain areas need to have major consideration in planning phases. For typical problems which will be encountered, see F. C. Murphy, *Regulating flood plain development,*

Dept. of Geog., Res. Paper 56 (Chicago: Univ. of Chicago Press, 1958); and G. F. White, ed., *Papers on flood problems,* Dept. of Geog., Res. Paper 70 (Chicago: Univ. of Chicago Press, 1961).
6. The Corps of Engineers has prepared lists of all communities above 2,500 population, summarizing the flood potential and status of engineering studies and/or projects for flood protection. For Iowa, see U.S. Army Corps of Engineers, *List of urban places, with information about flood problems, Iowa* (Rock Island: U.S. Corps of Eng., 1967).
7. See John R. Sheaffer, *Introduction to flood proofing, an outline of principles and methods* (Chicago: Univ. of Chicago Press, 1967).
8. See U.S. Army Corps of Engineers, *Guidelines for reducing flood damages* (Vicksburg: Corps of Engineers, 1967).

# 19

## THE IOWA FLOOD PLAIN MANAGEMENT PROGRAM

### James F. Cooper

THE FIRST COMPREHENSIVE REPORT prepared by the Iowa Natural Resources Council (INRC) which dealt directly with the principles of flood plain management was titled, "A Study of Flood Problems and Flood Plain Regulation, Iowa River and Local Tributaries at Iowa City, Iowa," and was completed in 1960.[1] Interestingly enough, the request from the Iowa City governing body for such a study was largely motivated by the initial rush to develop low-lying flood plain areas located downstream from the then recently completed Coralville flood control reservoir. Through (1) the delineation of the floodway and inundated areas and (2) the recommendation of minimum protection levels and the establishment of suggested encroachment limits, this report served as the technical basis for flood plain zoning at Iowa City.[2]

The continued and rapid encroachment on the flood plains of Iowa streams and rivers has led to increased emphasis on and a broadening of the state's flood plain management program. The coordina-

JAMES F. COOPER is Supervisor of the Flood Plain Management Section, Iowa Natural Resources Council, Des Moines, Iowa.

tion phase with the planning division of the Iowa Development Commission, as outlined in a previous chapter, has required further positive action by the INRC to develop a meaningful program.

In this chapter, the role of the INRC in carrying out the state's responsibilities in flood plain management will be described. Beginning with the initial Iowa City report, a broad foundation has been established that permits the local-state-federal relations in Iowa to function effectively. Through these relations, the INRC is endeavoring to achieve wise and prudent use of the flood plains in Iowa. The need for an active state coordination role will be all the more evident upon examining the factors discussed in this chapter.

## STATEWIDE OBJECTIVES

The present objectives of the flood plain management program of the INRC are basically no different than they were at the time of the preparation of the Iowa City report. These were to collect, analyze, develop, and distribute information relative to floods and flooding, whereby flood damages can be minimized and life, health, and property can be protected through planning, land-use control, and general flood plain regulation measures.

Specific objectives are:

1. To prepare reports which delineate and determine (a) the floodway, (b) the flood plain, and (c) minimum protection levels.
2. To promote the need for and merits of flood plain planning, land-use controls, and flood plain regulation.
3. To assist local governing bodies in flood plain planning, adoption of land-use controls, and regulations.
4. To assist the INRC in carrying out its statewide responsibility for the control, utilization, and protection of the water resources of Iowa, including flood control and regulation of flood plain construction.[3]

## SELECTION OF CRITERIA

Fundamental to delineation of flood plains and floodways and a determination of minimum flood protection levels is the establishment of criteria which will serve as the basis for uniform flood plain management programs conducted and coordinated on a statewide basis. As the first step, the physical limits of the flood plain need

# IOWA FLOOD PLAIN MANAGEMENT PROGRAM

to be defined by a large flood event which has some reasonable probability of occurring. Such a flood could be either (1) a flood resulting from probable maximum precipitation or (2) a flood equivalent to the standard project flood as conceived and used by the Corps of Engineers.[4]

The INRC has recognized that an extremely rare flood, largely hypothetical in determination, may not be a reasonable event upon which to base regulations or controls. Therefore, it has selected a flood of somewhat lesser magnitude, derived from experienced floods which have been observed throughout the state (excluding extremely rare events), as a design flood concept. This design flood is used for studying and establishing encroachment limits and minimum flood protection levels.[5]

In establishing encroachment limits, it is obvious that the more restrictive the encroachment, the greater will be the increase in stage for a given discharge. The general guideline in Iowa for establishing limits is to permit encroachment to setback limits located equidistant from the channel centerline, on opposite sides of a stream or watercourse, providing that the resultant increase in water surface elevation in and upstream of the reach involved does not exceed one foot. When limits are set, every reasonable effort is made to place existing development outside the floodway or area needed to convey floodflows.

Minimum protection levels are established by the elevation of the computed water surface profile corresponding to the design flood as affected by the selected degree of encroachment. This means that planning for future uses and construction of planned developments can proceed with no additional concern about adversely affecting others. Variances, however, would require additional review.

## Conducting Flood Plain Information Studies

In carrying out the objectives of the Iowa flood plain management program, based upon uniform statewide criteria, it is obvious that the whole state cannot be studied at once. Therefore, local units of government interested in being included in the priority of areas to be studied must presently apply through the INRC for either a flood plain information study by the Corps of Engineers or an encroachment study by the council, or for both. Coordination through the INRC is also desirable for flood plain studies undertaken by the U.S. Geological Survey (USGS), in cooperation with local units of government, if such studies may lead to the adoption of flood plain regulations and controls. To date, seven flood plain information studies by the Corps of Engineers and one encroachment study by the USGS have been completed or are in progress for Iowa communities and counties.[6]

In applying for a federal flood plain information study and/or an INRC encroachment study, an application is required which, along with a supporting resolution from the local governing body, indicates a definite interest in using the flood plain information once it becomes available. The declaration of intent has a bearing upon the establishment of priorities, as does the magnitude of the flood problem and rapidity of urban encroachment into the flood plain.

Information to be contained in the application is to cover the following points:

1. The length of the reach of stream for which the study is desired.
2. Purpose for requesting the study.
3. Use which will be made of the study upon completion.
4. Amount of basic engineering information already available.
5. The local official authorized to represent the applicant.
6. Legal representative of the applicant.

Upon submission of an application for a study, the request is considered by the INRC and, if approved, placed on a priority list. If the application is for a flood plain information study by the Corps of Engineers and has been approved by the INRC, the request is forwarded to the appropriate district office of the Corps with a suggested priority relative to other requests already on file. Generally, the priority of studies for which application has been received is based upon:

1. The availability of basic engineering information.
2. The extent and seriousness of the flood problem to be studied.
3. The apparent willingness of the local unit of government to make use of the flood plain information.
4. The date the application is received.

## ENGINEERING STUDIES

A considerable amount of physical data concerning the flood plain and main channel is required for an engineering study. Before the actual computation of water surface profiles can begin, the basic engineering information from which the computer input is drawn

## IOWA FLOOD PLAIN MANAGEMENT PROGRAM

must be available. In most cases, it is the responsibility of the local community to have or to secure the needed valley and channel cross sections and topographic maps prior to the actual initiation of a study. The INRC and the Corps of Engineers do not have the personnel or funds to secure extensive engineering information for these local studies. Because of this limitation it is strongly recommended that communities wishing to avail themselves of flood plain information or encroachment studies take the necessary steps to secure the needed topographic coverage. What can be more vital to various aspects of planning than an up-to-date physical representation of the area under consideration? Cooperative programs are available through the USGS for the statewide and national topographic mapping program.[7] However, more detailed, large-scale local surveys are usually required for the purposes of engineering studies for flood plain information reports.

After all the needed engineering information is assembled, the basic study approach is determined. For undeveloped areas, the study generally consists of determining the flood plain limits and the water surface profiles for the design flood. If the community which has requested the study is undecided as to the use or the degree of development to be permitted on the flood plains, several degrees of encroachment may be studied. This permits consideration of alternatives to arrive at a single degree of encroachment which will meet INRC criteria and will be acceptable to the community for planning and development purposes.

If the area is already in some stage of development, the study obviously becomes more complicated. Care must be taken in selecting encroachment limits so that large areas of existing development are not included in the floodway, which is defined as that area needed to convey floodflows. Encroachment limits must be transitioned into and out of bridge waterway openings, must consider existing street patterns, and of course must conform to any existing natural controls.

In highly urbanized flood plain areas, the floodway and corresponding encroachment limits often must be restricted to the existing channel area even though flood stages may be substantially increased by excluding overbank areas from the future development plan. Only where there is no other alternative is existing development included in the floodway area. Such a case would occur when the exclusion of existing development from the floodway would so reduce the effective area of floodwater conveyance that a severe restriction and a large increase in the upstream flood stage would result therefrom.

In an extensive reach of a stream where the channel itself is severely restricted, a large-scale channel-improvement plan can be studied. In such a case, the use of present-day electronic computer techniques permits the evaluation and determination of a so-called priority list for removal of restrictions in a reach-by-reach channel-improvement program.

Special mention should be made of bridges and road grades.[8] In Iowa it has been found that approach grades and bridges have a decided effect on the pattern of development in the surrounding area. Low approach grades promote adjoining low development and of course high approach grades just the opposite. Each situation can have its drawbacks, the first permitting overflow but inundation of the low-lying property, and the second creating a dam effect which in turn will create higher stages upstream. Of equal significance is the adequacy of the bridge opening itself. An inadequate opening causes higher flood stages upstream with the attendant tendency for overflow, both increasing the flood hazard to adjacent development.

Present engineering practice in Iowa is to design the bridge waterway opening to a 50-year flood frequency. There are instances when a bridge designed on the basis of a 50-year flood will create an effect upon existing areas which disrupts a predetermined plan of flood plain development by increasing upstream flood stages, thus reducing the level of safety against floods. Therefore, it is of paramount importance that the backwater effect of a proposed bridge structure on the design flood be determined and taken into account in computing the size of bridge waterway opening required, even though the size of the opening thus determined would exceed the normal 50-year flood frequency criteria.

The water surface profiles determined by the INRC are computed using an electronic digital computer. The computer methods are based on a backwater program initially prepared for the INRC through coordination with the Center for Industrial Research and Service of Iowa State University.[9] The program is written in Fortran computer language and will compute flow through bridges, over road grades, in main river channels, and across overbank areas. Although there may be more sophisticated programs, the one presently in use has two distinct advantages in that the input can be easily prepared and the output can be read and interpreted by personnel untrained in computer techniques.

The preceding discussion of the conduct of a flood plain study is based on the assumption that the entire study was made by the INRC. However, a federal agency in all probability will be con-

# IOWA FLOOD PLAIN MANAGEMENT PROGRAM

ducting one or more phases of many of the studies. The Corps of Engineers through its flood plain management services program determines water surface profiles and corresponding inundation limits for selected discharges under *existing* conditions.[10] The district offices of the Corps of Engineers working in Iowa have considered the computation of water surface profiles corresponding to *selected degrees of encroachment* during the course of preparing a standardized flood plain information study. Also, the USGS has completed a flood plain study which included the computation of water surface profiles corresponding to selected degrees of encroachment.

For sound comprehensive flood plain planning and effective flood plain land-use regulations and controls to be achieved, information relative to permissible degrees of encroachment as well as to the physical limits of the flood plain are essential. In Iowa the federal and state agencies working in the area of flood plain management have computer capabilities. It is highly desirable for one agency to make its computations and data available to the others. By doing this, duplication of effort will be avoided, and the results will necessarily avoid any nonuniformity or discontinuity.

In the final analysis, it is the responsibility of the INRC to interpret the results of any flood plain inundation and flood plain encroachment computations. The interpretation made will be based on established policy relative to encroachment limits and minimum protection levels which are directed insofar as possible to meet the needs and desires of the local community.

When a survey is conducted, close coordination is maintained with the local unit of government for the purpose of generally locating encroachment limits which will best suit the interests of the community. When computations are complete, any previously proposed plan is reviewed to ensure its compatibility with INRC requirements relative to its hydraulic effects upon flood stages and profiles. If compatible, the encroachment limits are accepted for report preparation.

Typical results of flood plain information studies are shown in Figure 19.1. As developed by the Corps of Engineers for Indian Creek and Dry Creek at Cedar Rapids, these figures illustrate in plan view the areal extent of inundation by floods of various degrees of severity.[11] The standard project flood identifies those areas which have a reasonable probability of becoming flooded sometime in the future. Flood profiles included in the reports provide elevation data for the selected flood discharges.

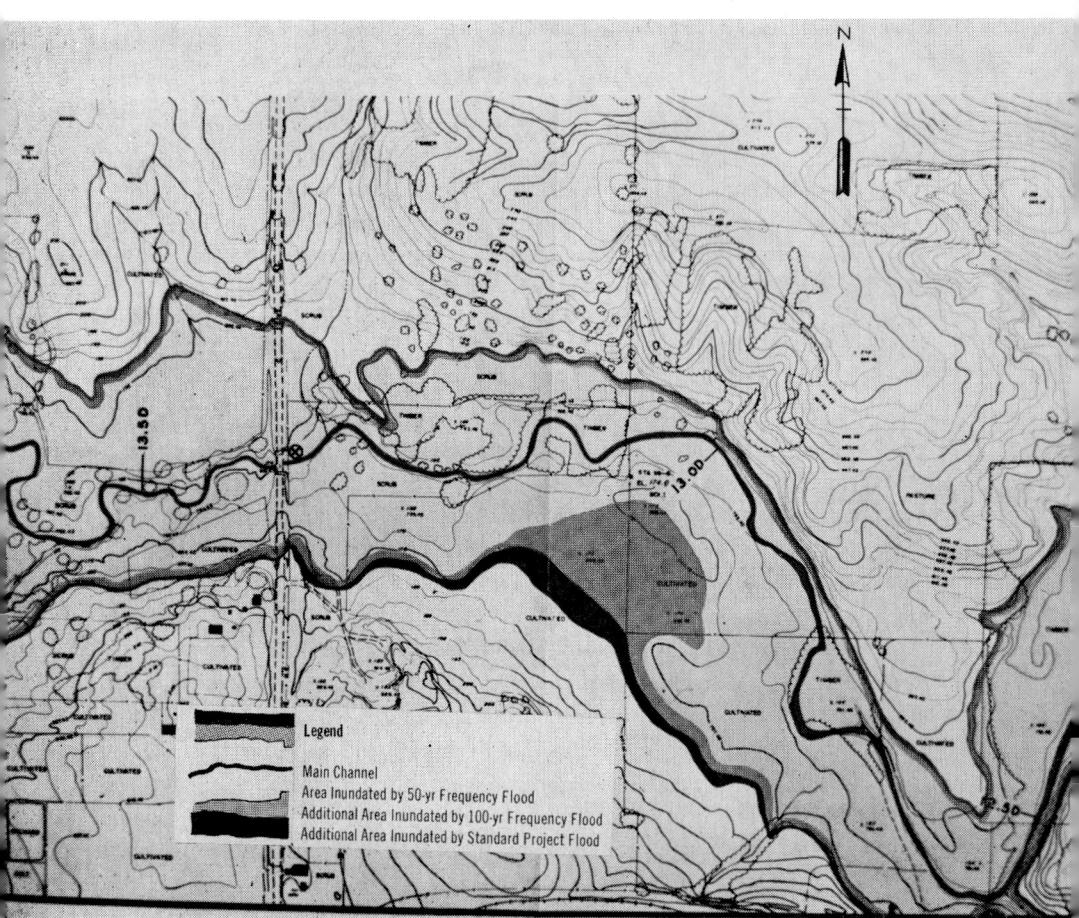

Fig. 19.1. Flood plain area-inundated map for Indian Creek and Dry Creek at Cedar Rapids, Iowa (after U.S. Army Corps of Engineers, 1965).

## THE FLOOD PLAIN MANAGEMENT SUMMARY REPORT

The stage is now set for preparation of a summary flood plain management report by the INRC.[12] It is intended that the report be in a form which will serve as a ready reference for administration and implementation of local regulations and controls by the community. Generally, the report will include:

1. Discussion of the design flood.
2. Delineation and discussion of flood plain limits, encroachment

# IOWA FLOOD PLAIN MANAGEMENT PROGRAM

limits, and water surface profiles corresponding to the design flood.
3. Discussion of existing natural and man-made controls on a reach-by-reach basis.
4. Discussion of minimum protection levels.
5. Coordination of information contained in the report with existing and proposed local ordinances, controls, and land-use programs.

Special mention should be made of the graphic form in which the flood plain and encroachment limits and water surface profiles for the design flood appear in the report. The profiles are drawn on profile paper to such a scale that the elevation of the design flood or the flood used to define the limits of the flood plain can be determined to a fraction of a foot at any point along the watercourse being studied. This is important because the elevations shown on the profiles will be the final authority in locating the limits of the flood plain and the physical limits of the minimum protection levels at any given location.

Delineation of the flood plains and encroachment limits are of equal importance. If at all possible, the limits of both are shown on topographic maps having a 2-foot contour interval. Current maps are used where available so as not to mislead anyone who views them. If the maps are not current, it is very carefully pointed out that all calculations reflect conditions at the time of the study, rather than conditions shown on the outdated maps. Although the elevations from the profiles ultimately determine the limits of the flood plain, the topographic maps are first observed and readily understood by the public. Any distortions or inaccuracies usually lead to public misunderstanding, confusion, and even distrust.

A sample of the results of a summary report prepared for the Indian Creek and Dry Creek area at Cedar Rapids is shown in Figure 19.2. The plan view shows the flood plain and encroachment limits delineated for flood plain regulation purposes.

Natural and man-made controls which affect flood stages are discussed in the reports to explain any variations in location of encroachment limits and variations in flood stages and depths. Such explanations will indicate to the community where channel improvements, bridge enlargements, and other remedial measures can best be undertaken to reduce flood stages and, consequently, flood damages.

As previously explained, the basis for establishment of minimum protection levels is the design flood. However, specific criteria for

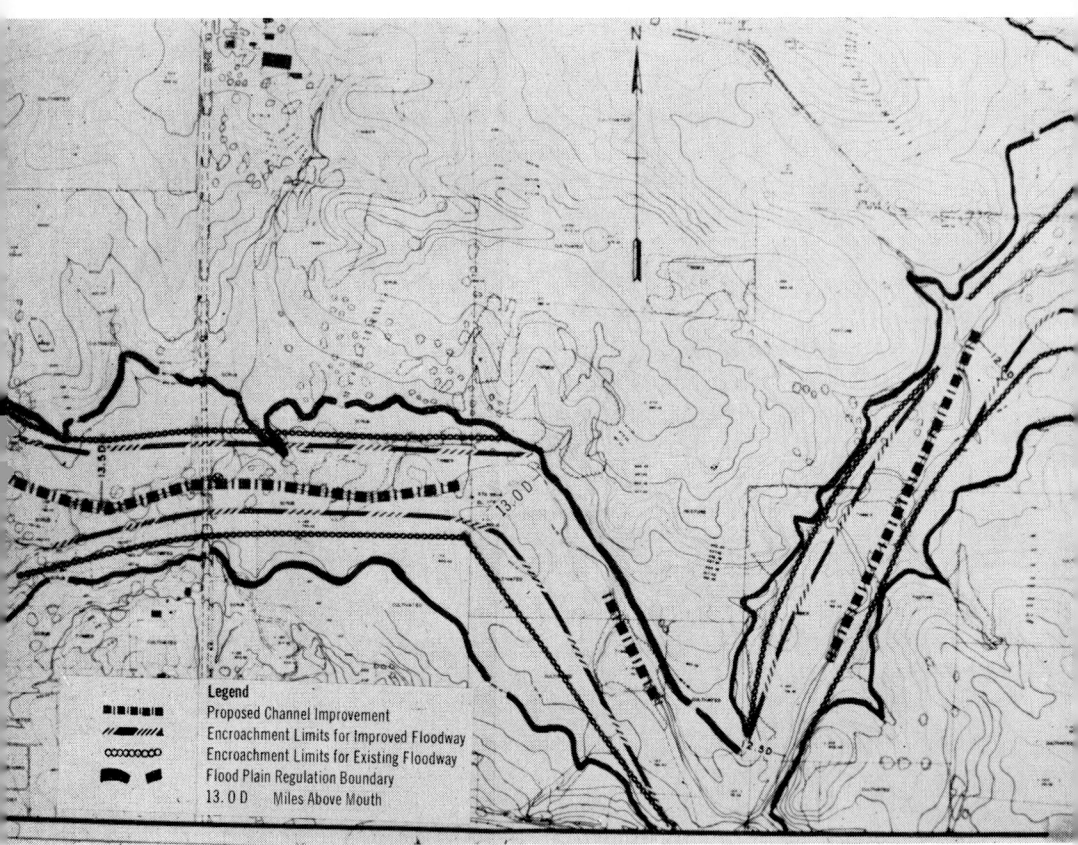

Fig. 19.2. Flood plain encroachment limits and zone boundaries for regulatory purposes, as proposed for Indian Creek and Dry Creek at Cedar Rapids, Iowa.

minimum protection levels relative to the design flood have been assigned to residential, commercial, and industrial types of development. Criteria have also been established for bridges, road grades, and other types of public utilities, as they would affect or be affected by the design flood.

If warranted, portions of the above criteria will be revised when construction and design specifications relative to floodproofing become available. Not only will such specifications be invaluable to architects and engineers engaged in design but they are needed by the cities, counties, and the INRC in passing judgment on the adequacy of the design of proposed floodproofing measures. Studies to formulate floodproofing specifications have been initiated at the federal level.[13]

## GAINING ACCEPTANCE OF REGULATORY MEASURES

After the report is completed and presented to the local unit of government, one and possibly two of the specific objectives of the Iowa flood plain management program have been achieved. Continued contact, coordination, and cooperation with the community is needed, so that the information contained in the report is put to the best possible use. If local units of government do not adopt flood plain regulations, the INRC has statutory authority to establish and enforce such regulations.[14]

Care must be exercised in the use of statutory powers. Rather than to "speak softly and carry a big stick," it is probably better to "speak convincingly and carry an invisible stick." It is far better to convince the public of the need to be regulated than to cause resentment by arbitrary action. To make the public want and even request regulation is to make them aware of the problems associated with flooding and how the community is directly and indirectly affected. To make the public aware, it is also necessary to make governmental bodies—and the planners, engineers, and architects working for them—aware of the severity of the flood problem.

So that these objectives might be achieved more easily, INRC has arranged with the Iowa Development Commission to review and comment on all planning studies undertaken through the federal urban planning assistance program.[15] Another means of making the public aware of the problem, of course, is to remain in contact with all the disciplines and agencies connected with local land use, planning, regulation, and control.

The procedural steps involved in the Iowa flood plain management program are summarized below. This summary illustrates the local-state-federal government role in achieving sound programs for development and use of the flood plains.

1. Local officials recognize the potential flood hazard from urbanization trends and unwise flood plain construction.
2. The community requests technical assistance through the INRC, which coordinates all federal-state-local flood plain studies.
3. Basic flood information and flood plain studies are provided through programs of federal agencies. Coordination is maintained

with the planning division of the Iowa Development Commission if community comprehensive planning is also involved.
4. The INRC uses basic flood data and makes additional studies concerning future land-use plans and considering various degrees of flood plain encroachment.
5. The community, with assistance of state agencies, selects the flood plain land-use plan which is most compatible with the goals of the community. Encroachment limits, limits of the floodway, and the limits of the flood plain landward of the floodway are determined.
6. Suggested flood plain and encroachment limits to be incorporated into state or local comprehensive flood plain controls are formally presented through a public hearing.
7. Flood plain and encroachment limits are adopted by the state and may also be adopted by local units of government.
8. If adopted by local units of government as well as by the state, the flood plain and encroachment limits may be included in zoning ordinances, subdivision regulations, and building codes, upon adoption by the local unit of government as well as the state.
9. The local unit of government administers the plan, regulates construction and use, issues building permits, and refers exceptions to the INRC.

## SUMMARY

The future of the Iowa flood plain management program will be affected by many things, including state legislative action, court action, and federal action including flood insurance, to mention only a few. Whatever philosophy and approach to flood plain management is followed in the future, it is anticipated that the general program will be vastly strengthened with the availability of additional hydrologic information and techniques of analysis in making engineering studies.

Through the use of the electronic computer, it now appears feasible to conduct rainfall, runoff, and hydrograph analysis on a comprehensive statewide basis. Extensive stream-channel flood routing now appears feasible. Furthermore, the INRC looks forward not only to adoption of a basic uniform method of determining flood flow frequencies by all federal agencies but also to the adoption of a uniform method of regional floodflow frequency correlation.

In conclusion, the Iowa flood plain management program is gaining momentum and the INRC will continue to look forward to

# IOWA FLOOD PLAIN MANAGEMENT PROGRAM

excellent working relationships with the county and city governments, various state agencies involved in planning for water resources, engineers, planners, architects and their groups, associations and societies, state educational institutions, and federal agencies working in flood plain management and water resources. Through the combined efforts of all these individuals and groups, a sound and continuing program can be achieved.

## REFERENCES

1. This initial study was made at the request of the Iowa City Plan and Zoning Commission and summarized in Iowa Natural Resources Council, *A study of the flood problems and flood plain regulation, Iowa River and local tributaries at Iowa City, Iowa,* mimeo., Des Moines, 1960.
2. Published originally in the Iowa City *Press-Citizen,* Aug. 7, 1962. See also Iowa City, City of, Zoning ordinance 2238, 1962; and a summary by J. W. Howe, Modern flood plain zoning ordinance adopted by Iowa City, *Civil Eng.* 33 (Apr. 1963): 38–39.
3. See Iowa Code, Ch. 455A, as amended 1966.
4. Evaluation of the probable maximum precipitation (PMP) is carried out by the hydrometeorological section of the U.S. Weather Bureau; and a recent summary and isopluvial maps are found in U.S. Weather Bureau, *Rainfall frequency atlas of the United States,* Tech. Paper 40 (Washington: USGPO, 1960). Values of PMP have been approached or equaled, unofficially at least, at several locations in the United States. See R. K. Linsley et al., *Applied hydrology* (New York: McGraw-Hill, 1949). The Corps of Engineers, to evaluate floods of lower magnitude than those obtained from PMP concepts, introduced the standard project flood. Although computed from regionally experienced storms, it roughly is 40 to 60 percent of the flood computed from PMP analysis in most instances. See U.S. Army Corps of Engineers, *Standard project flood determination,* Civil Works Eng. Bull. 52-B (Washington: Corps of Engineers, 1967).
5. The analysis and initial report of the design flood concept were reported by Eldred Rich, *Study of regional floods in Iowa,* unpubl. rept. presented at the County Engineers' Hydraulics Short Course, Iowa State Univ., Ames, Feb. 1966.
6. See Chapters 17 and 18, supra. Reports of the completed studies may be obtained from the INRC or the respective federal agency involved. Summary pamphlets also have been prepared for general distribution.
7. A summary of the state needs for topographic mapping of all areas of Iowa can be found in Iowa Natural Resources Council, *Report for the biennial period July 1, 1962, to June 30, 1964,* Des Moines, 1964.
8. See U.S. Bureau of Public Roads, *Hydraulics of bridge waterways,* HDS 1 (Washington: USGPO, 1960).
9. The complex nature of river hydraulics and the usefulness of detailed computer programs for evaluating encroachment limits led to the development of the program presently in use. See J. O. Shearman and M. D. Dougal, *A computer program for computing water surface profiles* (Ames: Center for Ind. Res. and Serv., 1964).

10. See U.S. Army Corps of Engineers, *Guidelines for reducing flood damages* (Vicksburg: Corps of Engineers, 1967).
11. Five flood plain information reports have been completed in Iowa by the Corps of Engineers for flood plain areas at Cedar Rapids, Davenport, and Ames. For a typical report, see *Flood plain information report, Indian and Dry Creeks, Linn County, Iowa* (Rock Island: U.S. Army Eng. Dist., 1964).
12. Two major encroachment studies have been completed, both being follow-up studies from previous Corps of Engineers flood plain information studies. See Iowa Natural Resources Council, *Effects of flood plain encroachment, Indian and Dry Creeks, Linn County, Iowa,* summary rept., mimeo., Des Moines, 1966; and Iowa Natural Resources Council, *Flood plain encroachment study, Duck Creek, Scott County, Iowa,* summary rept., mimeo., Des Moines, 1967.
13. Two reports are currently available: J. R. Sheaffer, *Flood proofing: An element in a flood damage reduction program,* Dept. of Geog., Res. Paper 65 (Chicago: Univ. of Chicago Press, 1960); and J. R. Sheaffer, *Introduction to flood proofing, an outline of principles and methods* (Chicago: Univ. of Chicago Press, 1967).
14. Iowa Code, § 455A.35, 1966.
15. See U.S. Dept. of Housing and Urban Development, Programs of HUD, IP-36 (Washington: USGPO), 1967.

East Nishnabotna River at Atlantic, 1958. Courtesy Corps of Engineers, Omaha District.

# Part 7

"... the benefits of the flood plain location are only those not obtainable at an alternate highland location."

R. K. Linsley and J. B. Franzini

# SUCCESS OF FLOOD PLAIN MANAGEMENT PROGRAMS IN OTHER STATES

# 20

## LINCOLN'S EXPERIENCE IN REGULATING FLOOD PLAIN DEVELOPMENT

### James H. Schroeder

LOCAL COMMUNITIES and metropolitan areas must take a substantial interest in and a positive stand toward programs for encouraging the wise use of flood plain lands. Public support is a first necessity, sound technical planning a second requirement, and the third factor of critical importance is the ability of community leaders and elected public officials to adopt and carry out the proposed programs. All these phases have been experienced in Lincoln, where concern for the flood problem led to community action. The experience gained by the Lincoln City–Lancaster County Planning Commission in flood plain regulation may be of value to others and will serve as the basis of this discussion.

### HISTORICAL INFORMATION

The city of Lincoln is located in Lancaster County in the southeastern part of Nebraska. It was first settled more than 100 years ago

JAMES H. SCHROEDER is Deputy Planning Director, Lincoln City–Lancaster County Planning Commission, Lincoln, Nebraska.

and has grown to approximately 155,000 residents, with another 20,000 people living in the county outside the city limits. The city covers 49 square miles in area. Almost all the urban growth has occurred to the east of Salt Creek, the significant development on the west side consisting of the Lincoln airport and some scattered residential developments. It should be noted that the city limits represent almost the full extent of urban growth; there is very little urban fringe outside the corporate limits.

The first settlement at Lincoln was made near the banks of Salt Creek, the principal stream in the Lincoln area. In fact, all but the southern 10 percent of Lancaster County lies within the Salt Creek watershed. This creek is a tributary of the Platte River, and the combined Salt-Wahoo watershed has a total drainage area of 1,665 square miles.

The growth of the city has occurred in a fan shape to the east of Salt Creek. One-half the 16 small tributary streams are in the immediate Lincoln area, and five of these have posed serious flooding problems to urban growth in the past.

Floods have caused distress and damage to the Lincoln area since settlement first began. The first flood on record occurred in 1861, followed by approximately 100 since that time. Seventeen of these have been termed major floods, 34 moderate, and the remainder minor. Two floods stand out especially. The first occurred in July, 1908, and the second in May, 1950. Seven people lost their lives in the 1908 flood, and nine people died as a result of the 1950 flood. As recently as 1963 three people lost their lives in floods in the Lincoln area.

As a result of these experiences the public was painfully aware of the flood problems. However, as with many of the problems in urban areas, the situation had to deteriorate to a critical level before efforts were initiated to solve them. In the past eight years much has been done in the form of corrective measures within the Lincoln area, and some accomplishments have been in the form of preventive measures.

It will be of value to discuss the preventive measures first and then to highlight some of the corrective steps that have been taken.

## FLOOD PLAIN PLANNING

Comprehensive plans developed for Lincoln in 1952 and again in 1960 have recognized the flood problem. The 1960 comprehensive plan pertains both to the city area and to the whole of Lancaster County.[2] It includes the usual elements of a land-use plan, a major street plan, a utilities plan, and a park plan. Both the 1952 and the 1960 plans included specific sections in which the flood problem was studied. The completed plans provide for limiting the development to be allowed in the flood plain area. This is done in two ways:

1. Lands which lie in flood plains are designated to be a part of the community's park system. This is provided for in the land-use plan and in the more detailed proposed park system.
2. Where it is not possible to include the land in the park system, the development of the land within the flood plain is to be kept at an absolute minimum. This concept is spelled out in the text of the plan.

This latter course of action can only be done if additional actions are taken by the city—or, in the case of municipal utilities, if certain actions are not taken by the city.

## CONTROL OF UTILITY EXTENSIONS

It is generally recognized that for intensive urban development to occur, the extension of utilities into an area is required. The city of Lincoln has made a concerted effort to prevent this from happening by controlling the expansion of municipal utilities.

Resolutions adopted by the Lincoln City Council have limited the extension of utilities into areas subject to flood. This drastic but effective control measure can be justified if we realize how the extension of utilities encourages growth. Regardless of whether such extensions are in a flood plain or some other part of the city, their availability encourages growth in that direction, just as do the construction of a new school, the improvement of a traffic artery, or other similar facility.

However, several things had to happen before the city council could realistically adopt such a resolution. First, it was necessary to know more precisely the location of the areas subject to flood, and second, what flood magnitude was to be used for regulatory purposes.

After receiving sufficient support from the community and the different governmental subdivisions in the area, the Lincoln City–Lancaster County Planning Commission requested the U.S. Army Corps of Engineers to conduct flood plain information studies on Salt Creek and its tributaries in the Lincoln area. Significant support was generated through the Salt Valley watershed district.[1]

This request was made in 1962, and the technical studies began in 1963. The first of three volumes with the results of these studies was transmitted to the city in 1965. Salt Creek and two of its tributaries were included in this first report. Municipal personnel participated in the studies by providing some of the necessary survey information and by reviewing preliminary drafts. The last of the three volumes was presented to the city in September, 1967.

The information obtained from these studies is of the following type:

1. The areas subject to inundation from a 100-year flood and lesser floods of the 50-, 25-, and 5-year magnitude and frequency are shown.
2. The stage of each of these floods is identified, to indicate the depth to which flooding will occur.
3. The principal land forms and man-made facilities which act as barriers to the movement of water are also identified.

This information is presented in map form, outlining the areas subject to flood, and in the form of stream profiles and cross sections of selected locations on each stream.

## Regulating Utility Extensions

After having received Volume I of the flood plain reports and reviewing its contents, the city council adopted a resolution in October, 1965, containing several important provisions:

1. The resolution pertains only to the streams for which information is shown in Volume I.
2. It includes the recommendation of the Lincoln City–Lancaster County Planning Commission that urban development be limited in flood plains.
3. The wording of the resolution recognizes that when the city uses its credit and governmental services to extend public facilities into areas subject to flood, it encourages the citizens of the city to believe that the area is safe from flooding and, further, that it is not in the public interest to encourage urban development in areas subject to flooding.
4. The resolution identifies the flood area as that designated in the report as being subject to inundation by the 100-year flood.

## LINCOLN'S EXPERIENCE

There are three very important features of this resolution. First, it recognizes that it is not in the public interest to encourage urban development in areas subject to flooding. Such development creates the potential for hardship and expense, both to those living or working in the flood plain development and to the community as a whole. Second, it states that no private connections may be made to utilities lying within the flood plain. This can act to discourage the continuance of development already in the area. Third, it provides that when utilities have to be extended through a flood area to reach other areas the flood area is not to be assessed for any special benefit derived from the utility extension. This reduces the pressure to allow development to encroach on the flood area. This last point illustrates the need to incorporate consistency into land assessment policy. It does not make sense to pass ordinances or resolutions unless assessment practices are consistent with the intent of the resolutions or ordinances.

The first resolution affected only those streams covered by Volume I—Salt Creek, Beals Slough, and Haines Branch. In September, 1967, the city council passed a similar resolution covering those streams included in Volume II of the Corps of Engineers study. A resolution covering the streams in Volume III is being prepared for consideration by the city council.

### Success of the Regulation Measure

How well has this resolution worked? To date, we have not really had a test of the council's action. Development has shied away from these areas, so that the city has yet to be challenged in the context of this resolution.

Unfortunately, there has been some urban development in flood plain areas. Of recent significance is a high-cost housing project on Salt Lake at the west edge of the city. This was constructed in an area originally thought not to be within the flood plain. However, Volume III does identify the area of inundation for the 100-year flood along Oak Creek as extending into this residential and lake area. The developer is concerned about the possibility of flooding and is working with the city and the Corps of Engineers in an attempt to reach a solution. There are fairly substantial and widespread areas lying within the flood plain in older sections of the city, and a comprehensive flood plain management program which can control and regulate these areas has not yet been devised.

## ALTERNATIVE CONTROL MEASURES

The first preventive measure mentioned, that of designating in the land-use plan that land within flood plains be a part of the community's park system, continues to play a large role in the program at Lincoln.

Figure 20.1 shows the proposed park system as extracted from the comprehensive plan. Also shown is the existing system, that is, as it existed in 1960. Extensive areas along the streams have been designated as a part of the future park system. By setting these areas aside as open space or land which is not to be developed, the city hopes to reduce the cost of necessary flood control programs and the amount of potential flood damage. A moderate degree of success has been achieved with this plan.

The areas outlined in Figure 20.1 are those which have been acquired for public open space since 1960. Progress also has been made in acquiring all the land to the south along Salt Creek. There are approximately 1,300 acres of land along this stream, acquisition of which is proposed for use as a "wilderness park."

This acquisition involves the cooperation of the county, the city, and also the Salt Valley watershed district. The county is acquiring the land with the aid of an open-space grant from the Department of Housing and Urban Development.[3] An agreement has been reached wherein the city will develop the land and maintain it; development in the park will be only minimal. The Salt Valley watershed district is participating in the cost of acquisition and will retain an easement over the property for flood control purposes. Some of the county's funds for acquiring this land will come from private foundations in the Lincoln area.

Approval has been received of the first open-space grant for acquisition of about half the area, and the actual purchase of the lands has begun. The second application for a grant on the remaining land should be ready to submit in the near future. This particular program represents a substantial accomplishment both for the flood control program and for the parkland program.

One combined facility, a flood control structure and park area, should be singled out for discussion. This project has been a significant accomplishment in the Lincoln program. The structure and park were originally proposed in the 1952 comprehensive plan. The development of this plan again represents close coordination and cooperation between different governmental bodies—city, county, Salt Valley watershed district, and the Corps of Engineers. The multi-

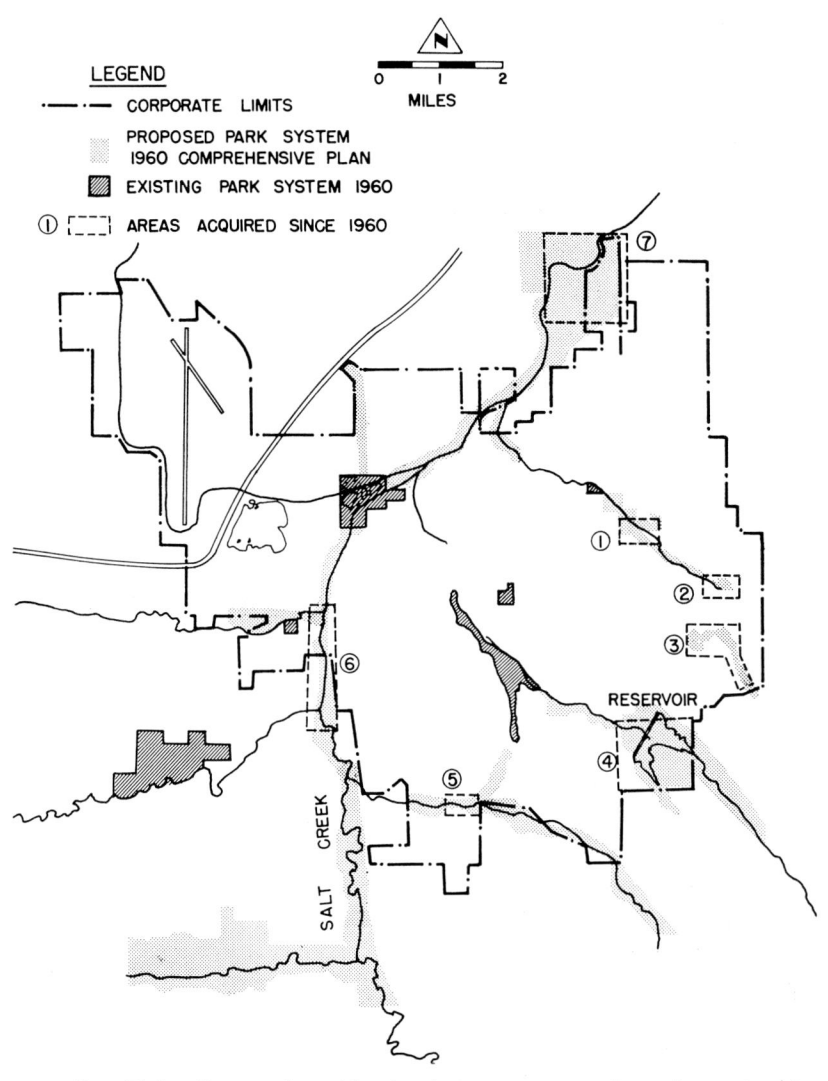

Fig. 20.1. Proposed parklands along streams and creeks at Lincoln, Nebraska.

purpose facility makes optimum use of the flood plain and valley area in this part of the metropolitan area.

A second area, acquired at the extreme north end of the city, represents another example of the multiple use of land. It has been purchased through an open-space grant for park purposes; however, it is being used initially as a sanitary landfill by the Department of

Public Works. The Parks and Recreation Department will prepare a development plan to be carried out in the future.

Miscellaneous land areas along streams also have been acquired, and others have been set aside through private open-space development. However, success in the parkland program along streams has been limited. The city has only limited funds for the acquisition, development, and maintenance of parks as well as all other services and facilities. Therefore, some other areas and programs may have higher priority, and frequently acquisition of flood plain lands for park purposes is delayed or postponed.

## SUMMARY

The procedures outlined above represent the extent of efforts for preventive measures in the field of flood control and flood plain regulation. A need for additional controls in this area is recognized; however, zoning may not necessarily be the most appropriate route to follow. Perhaps a combination of measures is required. Several methods discussed in this book cover the whole range of recommended procedures, and it cannot be expected that one can accomplish the task alone.

Because of a rather unique position, Lincoln is in a better position to control all development than are most other communities. Complete zoning and subdivision control, both within the city and in the area three miles from the city limits, is in force. Building permits also are issued by the city for this area. In addition, an ordinance defines a subdivision as any division of land which creates one or more parcels of less than five acres in size. Since the city approves all subdivisions, the problem does not exist whereby individuals sell off one, two, or three small parcels by a metes and bounds description to escape regulation measures.

The development of a more comprehensive flood plain management program is needed, and—of special importance—it should be integrated into the comprehensive plan for the community.

One additional aspect of water resources management needs to be mentioned. Substantial progress has been made in corrective flood control measures in the Lincoln area. Most of this is the result of work of the Salt Valley watershed district,[1] which was formed in 1960.

First, a land-treatment program was initiated in which the Soil Conservation Service and local soil and water conservation districts assisted individual farmers in contour farming, terracing, establishing grassed waterways, and other good farming practices. Next came the construction of farm ponds and gully control structures which impound the excess water and sediment that run off the fields and pastures despite wise conservation practices. The third segment of the program called for the construction of smaller flood detention dams to store water during the flooding periods, while permitting some of the higher land behind the dams to be farmed during the dry periods. Also included in the flood control program are nine large dams constructed by the Corps of Engineers and the construction of needed channel improvements along Salt Creek and some of its tributaries. These dams are also being developed as much-needed recreation facilities by the Nebraska Game Commission, an excellent example of joint use.

Much of this work has been completed. The fact that it has been beneficial can be attested to by reviewing the effects of the extremely heavy rains which occurred in June, 1967. The waters either were held in the many retention dams and ponds around the city or were channeled through the Lincoln area with only a minimum of flooding and damage. It is estimated that $1 million of flood damage losses were prevented as a result of the flood control work that had been completed.

In the Lincoln area both the public and the city officials are aware of the additional work required to conserve water and control flooding. More construction and maintenance work is needed, and further preventive measures must be developed. These will be assisted by recent legislation at the state level. Legislative Bill 893 passed by the Nebraska Unicameral in 1967 makes it mandatory that all political subdivisions in the state adopt land-use regulations to control development within flood areas. This act should be of benefit as obligations continue to be met with respect to flood plain problems. Perhaps past actions do not represent anything like a comprehensive approach, but the city of Lincoln hopes to continue improving the program—a challenging assignment.

## REFERENCES

1. Geier, L. *Valley of still waters, the story of the Salt-Wahoo watershed.* Lincoln: Salt-Wahoo Watershed Association. Ca. 1960.
2. Lincoln City–Lancaster County Planning Commission. *Comprehensive regional plan, for the Lincoln City–Lancaster County metropolitan area of Nebraska.* Lincoln. 1960.
3. U.S. Department of Housing and Urban Development. *Programs of HUD.* IP-36. Washington: USGPO. 1967.

# 21

## EXPERIENCE OF THE TENNESSEE VALLEY AUTHORITY IN LOCAL FLOOD RELATIONS

John W. Weathers

THE PREVIOUS CHAPTERS have been devoted primarily to the technical and legal aspects of flood plain management. Current problems have been used frequently in illustrating the techniques which are being proposed or adopted to solve flood plain management problems. As flood plain managers wrestle with the trials and tribulations inherent in regulating land uses, programs which are being tried elsewhere are of substantial interest and importance. Through these various measures they are endeavoring to promote an effective program for flood plain management, which will provide for the optimum local use of flood plain lands.

Therefore, it is appropriate that this final chapter be directed to an appraisal of management programs which have developed through long-term experience in other regions of the nation. In this manner the potential effectiveness of proposed programs for local

JOHN W. WEATHERS is Chief of Local Flood Relations, Tennessee Valley Authority, Knoxville, Tennessee.

communities may be evaluated more readily. The purpose of this chapter is to examine the experience record of the Tennessee Valley Authority (TVA), which has been active in the field of water resources since the 1930's, and in the specific area of flood plain management for 14 years. Through concerted and dedicated efforts, it has attained a reasonable measure of success.

This approach will, perhaps, give the reader a renewed spirit of enthusiasm. One additional ray of hope—this summary, devoted to a general review of TVA experience, is written by one who has experienced and overcome frustration. Therefore, there is reason to be less a "prophet of doom" and thus to proceed constructively toward finding solutions to our flood plain management problems.

## WATER RESOURCES DEVELOPMENT BY THE TVA

To provide perspective for subsequent comments and discussion, background information about the Tennessee Valley Authority is needed. The TVA is a corporate agency of the federal government created by an act of Congress in 1933 and operates somewhat as a private corporation.[1] It is charged with the broad duty of planning and effecting programs for the development of the resources of the Tennessee River drainage area. The Tennessee River basin contains 41,000 square miles. Within the watershed are parts of seven states, including all or parts of 125 counties. The topography varies from mountainous regions to gently rolling hills and plains. The average annual precipitation is 51.5 inches.

The system of TVA dams and reservoirs on the main stream and its tributaries makes the Tennessee River the best-controlled major stream in the world.[2] However, even this great system of engineering works does not provide complete flood protection for communities along the river and gives no protection to many communities located along uncontrolled tributaries.

There are nearly 150 urban areas in the valley which can suffer severe damage from floods. Conferences among TVA engineers and state planners within the Tennessee River basin culminated in agreement that these communities could and should take action to alleviate their flood problems. To do this they needed technical information on the flood hazard and technical guidance and assistance in developing plans. The local flood relations program was established by the TVA over 14 years ago to provide such information and assistance.[3]

EXPERIENCE OF THE TVA 249

## THE TVA LOCAL FLOOD RELATIONS PROGRAM

Since its inception TVA's method has been to merge itself and its efforts with the initiative and institutions of the region,[4] insisting that flood damage prevention planning must be a cooperative federal-state-local venture. The elements of such a plan for a comprehensive flood damage prevention program are shown in Figure 21.1. The cooperative program involves four general types of activities:

1. Reports outlining local flood conditions.
2. Flood plain regulations.
3. Spot assistance with flood problems.
4. Comprehensive flood damage prevention projects.

Altogether, a total of 99 local flood reports covering 120 communities have now been issued through their state planning agencies at the request of these communities. These reports are quite similar in content and format to the flood plain information reports of the U.S. Army Corps of Engineers.[5] The TVA does not have a special section to make these studies and prepare the reports. Rather, it utilizes the training and experience of the engineers and technicians in

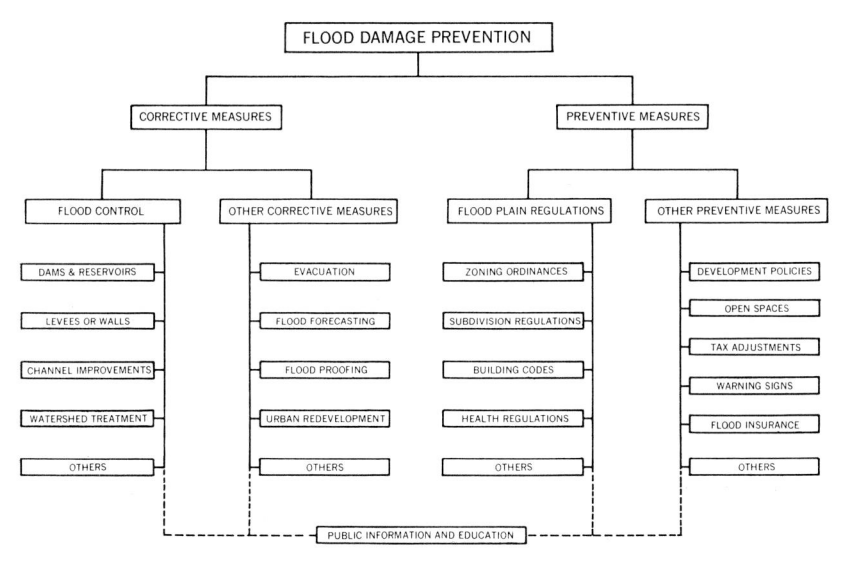

Fig. 21.1. Elements of the flood damage prevention program as developed by the Tennessee Valley Authority.

the division of Water Control Planning. In the peak production year, 12 reports were issued. Currently, only three or four per year are being prepared.

Early in the program, members of the TVA staff and state planners met with local officials to convince them that they needed factual flood data. Within a short time the situation changed, and requests exceeded the capacity to produce. Priorities were established to meet the most urgent needs as supported by the planners.

The reports have been widely used by government officials, planners, engineers, architects, industrial agents, developers, and others. Requests for many of the reports have required a second or third printing. Officials of two cities requested every copy of their flood reports that were published—so that they could burn them! Further contacts with these individuals disclosed that they were willing to destroy facts which might jeopardize their chances of getting an industry to locate in their community. Fortunately, this type of opposition is rare, and even these two cities have now accepted the reports and used the flood data.

### Implementation of Flood Plain Regulations

Following completion of a flood study, the TVA works with the community in preparing appropriate regulations for the control of uses in the flood plain. At this point the program becomes truly cooperative.[6] Land-use regulations are normally recommended to the city governing body by the local planning commission. Technical planning assistance is provided to the commission by a permanent staff, a private consultant, or the state. Engineering assistance, particularly in evaluating floodways, is provided by the TVA. The end results are a zoning ordinance and subdivision regulations which are acceptable to engineers, planners, and local officials. Final adoption and enforcement are a local responsibility.

To date, 56 communities within the Tennessee Valley have adopted zoning ordinances or subdivision regulations, or both, which contain flood plain provisions, and others are considering such action. Experience shows clearly that *the critical point is approval by the local planning commission.* These citizens must be convinced that the flood problem is serious enough to justify use of the police power and that the particular regulations proposed are fair and reasonable.

Flood plain regulations are most often attacked with the argument that they constitute a taking of land without due compensation. Extreme care must be taken to ensure that all uses of the flood plains are permitted which will be compatible with the intent of the regulations.

There are also those individuals who oppose any restrictions on the use of their lands. They will insist—often with vigor—that they should be permitted to take the risk and bear any damages. For about 10 years the planning commission in one valley town has periodically proposed flood plain zoning. Each effort has been defeated by the opposition of one man who is a subdivision developer. Paradoxically, he has strictly adhered to the recommended regulations in planning his developments. This man is willing to accept advice but resents legislation.

**Additional Coordination Efforts**

Flood reports published by the TVA do not cover all the areas within the valley where flooding might occur. Efforts have been directed primarily to existing urban areas which have experienced flood damages. With increasing population and improved transportation routes, developments are being constructed in areas which were previously open space. These may be residences, factories, shopping centers, schools, churches, utilities, civic buildings, or any of the facilities formerly located within the corporate limits of the cities. Each year the TVA receives about 70 requests for spot flood information for such sites. Providing this assistance has contributed significantly to reducing potential flood damages.

Flood information has been used in revising plans or locations for several schools which would otherwise have been vulnerable to flooding. Industries and commercial enterprises have revised their site plans, so that buildings would be on the higher ground and lower areas would be used for parking. In at least two known cases, shopping center developers have relocated the principal buildings and enlarged the stream channels to remove or reduce the flood threat. Federal agencies have used flood data extensively in processing applications for loans, grants, and mortgage insurance.

At Lewisburg, Tennessee, the Federal Housing Administration did not approve two subdivisions because of the flood hazard. This action was resented by the developers; but within five months after the time that homes would have been built, a flood occurred that was

4 feet above planned floor levels. These and similar occurrences have made the TVA confident that benefits have far exceeded the costs of producing the flood data.

### Role of Engineering Works

The TVA program does not rule out the possibility of constructing such flood control facilities as dams and reservoirs, channel improvements, and levees. For some communities within the valley a brief review of the flood situation indicates that a degree of flood control is desirable and may be feasible. Planners and local officials for these communities are encouraged to consider comprehensive solutions, using engineering and other technical assistance provided by the TVA.

Several communities are using this approach. Each appoints a flood study committee to consider the various ways by which flood damages might be prevented—floodproofing of buildings, flood plain regulations, flood forecasting, evacuation, opportunities created by urban renewal and highway relocation, and construction of flood control works. The committee's plan is submitted to the governing body for approval.

Where flood control facilities appear to be an essential part of the solution, the TVA investigates alternative plans, compares benefits and costs of each, and prepares a report on the most suitable. This report, plus a costsharing agreement, is used in seeking congressional appropriations for the federal share of the cost. When funds are made available, the TVA proceeds with design and construction of the facilities.

## CONCLUSIONS

The local flood relations program of the TVA has demonstrated that flood damages can be reduced to a reasonable level by cooperative action. An effective federal-state-local program permits each level of action to complement the others. Accurate flood data can be provided, planning alternatives can be outlined, and local plans for use and regulation can be adopted and enacted with adequate support by an informed public. The TVA program has also shown clearly that most communities will need continuing assistance after flood information is made available to them. Technical people furnishing this assistance must be able to overcome disappointments and frustrations which occur when concerted attempts to gain public support and local action occasionally fail. As more and more communities adopt flood plain management programs, the satisfaction of knowing that people will be free from the devastation of floods is more than sufficient reward.

## REFERENCES

1. The concepts and effectiveness of the TVA program have been discussed and reviewed widely. For a comprehensive summary, see the President's Water Resources Policy Commission, *A water policy for the American people,* vol. 1; *Ten rivers in America's future,* vol. 2; *Water resources law,* vol. 3 (Washington: USGPO, 1950).
2. For the engineering principles and concepts of multipurpose development of the water resources of a river basin, see R. K. Linsley and J. B. Franzini, *Water resources engineering* (New York: McGraw-Hill, 1964). See also note 1, supra.
3. The success of the initial program and recommendations for extending it nationwide are found in U.S. Senate, Select Committee on National Water Resources, *Flood problems and management in the Tennessee River basin,* Committee Print 16, 86th Cong., 1st sess., 1959.
4. The TVA coordination concepts are reviewed in J. E. Goddard, The cooperative program in the Tennessee Valley, in G. F. White, ed., *papers on flood problems,* Dept. of Geog., Res. Paper 70 (Chicago: Univ. of Chicago Press, 1961).
5. See Tennessee Valley Authority, *Flood damage prevention, and indexed bibliography,* 5th edit. Knoxville, 1967. For a typical report on flood problems in TVA communities, see Tennessee Valley Authority, *Floods on Big Rock Creek in vicinity of Lewisburg, Tennessee,* Knoxville, 1954.
6. The TVA has also encouraged and supported additional planning and research studies by state and local agencies. See Tennessee State Planning Commission, *Planning for flood damage prevention at Lewisburg, Tennessee,* TSPC Publ. 277, Nashville, 1956; and H. V. Miller, *Flood damage prevention for Tennessee,* TSPC Publ. 309, Nashville, 1960. See also H. F. Morse, *Role of the states in guiding land use in flood plains,* Spec. Rept. 38 (Atlanta: Georgia Inst. Tech., 1962).

# APPENDIX

## LIST OF URBAN PLACES IN IOWA
### WITH INFORMATION ABOUT FLOOD PROBLEMS
Prepared by Corps of Engineers, U.S. Army

THE PURPOSE of this list is to alert all federal, state, and local agencies to the flood problem in the urban places throughout the nation and to inform them of the source of more information and assistance. It comprises all incorporated and unincorporated places having 2,500 inhabitants or more in 1960 and those towns, townships, and counties classified as "urban" in the 1960 census of population.

The list contains information about the flood problems, studies made, protection works, maps, flood warning systems, and sources of other data. With regard to protection works, it should be noted that seldom do they provide complete protection. Only those places protected against the standard project flood or in the Tennessee Valley against the maximum probable flood are essentially free from flooding. In many places, flood protection works reduce damages on only one of the streams affecting the community. In other instances a dam upstream provides only partial protection because of tributary inflow below it.

Maps are considered adequate only where their scales are 1" = 2000' or larger. The flood warning column received a "Yes" if the Weather Bureau considered the existing system adequate.

The office shown as the information source will either supply information needed or refer users to proper agencies who have more complete information. Additional copies of this list may also be obtained from these sources.

This list was prepared by the Corps of Engineers, U.S. Army, with the assistance of Soil Conservation Service, Department of Agriculture; Geological Survey, Department of Interior; Weather Bureau and Coast and Geodetic Survey of the Environmental Science Services Administration; and Department of Commerce. In some instances states provided information.

# EXPLANATION OF SYMBOLS

## Status of Flood Studies

| | |
|---|---|
| CSR | Corps survey report |
| CFPI | Corps flood plain information report |
| SCS | SCS watershed survey |
| GS | GS flood hazard report |
| TVAS | TVA flood study |
| TVAF | TVA flood hazard report |
| BR | Reclamation project survey containing flood hazard information |
| S | State survey |
| L | Locally accomplished survey including flood hazard reports developed with federal financial assistance |
| a | Study complete |
| b | Study active |
| c | Corps study authorized or SCS study application received, neither funded |
| d | No current study authorized or application received |

## Status of Flood Protection

| | |
|---|---|
| CE | Corps of Engineers project |
| SCS | Soil Conservation Service project |
| BR | Bureau of Reclamation project |
| TVA | TVA project |
| S | State constructed project (providing significant degree of protection) |
| L | Locally constructed project (providing significant degree of protection) |
| a | Project constructed (either local protection, reservoir, or combination; degree of protection undefined) |
| b | Project authorized, under construction (includes advance engineering and design) |
| c | Project authorized, not funded for construction or for advance engineering and design |
| d | Favorable project survey awaiting congressional authorization |
| e | Project found economically or engineering unfeasible; date (year) |
| 1 | Flood plain regulation adopted |
| 2 | Status of flood plain regulation unknown |
| * | Project found economically feasible but not accepted by local interest. Restudy currently under way at request of local interests |
| ** | Project found economically unfeasible, restudy currently under way at request of local interests |

*Addresses of Information Sources:*

U.S. Army Engineer District, Rock Island, Clock Tower Building, Rock Island, Illinois 61201.
U.S. Army Engineer District, Omaha, 6012 U.S. Post Office & Court House, 215 North 17th Street, Omaha, Nebraska 68101.
U.S. Army Engineer District, Kansas City, 700 Federal Building, Kansas City, Missouri 64016.
U.S. Army Engineer District, St. Paul, 1217 U.S. Post Office & Customhouse, 180 E. Kellogg Blvd., St. Paul, Minnesota 55101.

## STATE OF IOWA

List of Urban Places With Information About Flood Problems

| Place | Type of Flood Problem | Studies | Protection | Maps | Flood Warning | Information Sources |
|---|---|---|---|---|---|---|
| Albia | None | L b | | No | No | Rock Island |
| Algona | Stream overflow | CSR b | | No | Yes | Rock Island |
| Ames | Stream overflow, local drainage | CSR b CFPI a S b | CE b 2 | Yes | Yes | |
| Anamosa | Stream overflow | L b | | No | Yes | Rock Island |
| Ankeny | None | L b | | Yes | No | Rock Island |
| Atlantic | Stream overflow | CSR ab L b | CE e 43 2 | No | Yes | Omaha |
| Audubon | Stream overflow | CSR ab | CE e 43 2 | No | Yes | Omaha |
| Belle Plaine | None | L b | | No | No | Rock Island |
| Belmond | Stream overflow | CSR ab L b | CE e 65 2 | No | No | Rock Island |
| Bettendorf | Stream overflow, local drainage | CSR ab CFPI a S b | CE e 62 *2 L a | Yes | Yes | Rock Island |
| Bloomfield | None | L b | | No | No | Rock Island |
| Boone | None | L b | | Yes | No | Rock Island |
| Burlington | Stream overflow | CSR ab | CE e 62 ** 2 | Yes | Yes | Rock Island |
| Carroll | Stream overflow | L b | | No | No | Rock Island |
| Cedar Falls | Stream overflow | L b CSR b L b | 2 2 | No | Yes | Rock Island |
| Cedar Rapids | Stream overflow | CSR ab CFPI ab S ab | CE e 65 ** 2 L a | Yes | Yes | Rock Island |
| Centerville | Stream overflow | L b CSR a | CE b 2 | No | Yes | Kansas City |
| Chariton | Stream overflow, local drainage | | 2 | No | No | Kansas City |

| City | Flood source | Code | | | | District |
|---|---|---|---|---|---|---|
| Charles City | Stream overflow | CSR b<br>S b<br>L b | | 2 | Yes | Rock Island |
| Cherokee | Stream overflow, local drainage | CSR b<br>L b | | 2 | No | Omaha |
| Clarinda | Stream overflow | CSR c<br>L b | | 2 | No | Kansas City |
| Clarion | None | L b | | | No | Rock Island |
| Clear Lake | Stream overflow | L b | | | No | Rock Island |
| Clinton | Stream overflow | CSR ab | CE e 62 ** | 2 | Yes | Rock Island |
| Council Bluffs | Stream overflow, local drainage | CSR a<br>L b | L a<br>CE b | 2 | Yes | Omaha |
| Cresco | None | L b | | | No | Rock Island |
| Creston | Stream overflow, local drainage | L b | | 2 | No | Kansas City |
| Davenport | Stream overflow, local drainage | CSR ab<br>CFPI a<br>S b<br>L b | CE e 62 ** | 1 | Yes | Rock Island |
| Decorah | Stream overflow, local drainage | CSR a | CE a | 2 | No | St. Paul |
| Denison | Stream overflow, local drainage | L b<br>CSR b<br>L b | CE e 63<br>CE c | 2 | No | Omaha |
| Des Moines | Stream overflow, local drainage | CSR ab<br>GS a | CE b<br>L a | 2 | Yes | Rock Island |
| De Witt | None | S b<br>L b<br>L b<br>CSR a | CE b | | Yes | Rock Island |
| Dubuque | Stream overflow | L b | L a | 2 | Yes | Rock Island |

|  |  | Status of Flood |  |  |  |  |
|---|---|---|---|---|---|---|
| Place | Type of Flood Problem | Studies | Protection | Maps | Flood Warning | Information Sources |
| Dyersville | Stream overflow | CSR a | CE e 62    2 | Yes | Yes | Rock Island |
| Eagle Grove | None | L b |  | No | No | Rock Island |
| Eldora | None | L b |  | No | No | Rock Island |
| Emmetsburg | Stream overflow | CSR b |           2 | No | No | Rock Island |
| Estherville | Stream overflow | CSR b<br>L b |           2 | No | Yes | Rock Island |
| Evansdale | Stream overflow, local drainage | CSR ab<br>L b<br>L b | CE e 65 * 2<br>L a | Yes | Yes | Rock Island |
| Fairfield | None |  |  | No | No | Rock Island |
| Forest City | Stream overflow | L b |  | No | No | Rock Island |
| Fort Dodge | Stream overflow | CSR ab | CE e 57    2 | No | Yes | Rock Island |
| Fort Madison | Stream overflow | CSR ab<br>L b | CE e 62 ** 2 | Yes | Yes | Rock Island |
| Glenwood | Stream overflow | L b |           2 | Yes | Yes | Omaha |
| Grinnell | None | L b |  | No | No | Rock Island |
| Hampton | Stream overflow | L b |  | No | No | Rock Island |
| Harlan | Stream overflow, local drainage | CSR ab | CE e 43    2 | No | Yes | Omaha |
| Hawarden | Stream overflow, local drainage | L b<br>CSR a | CE a       2 | No | Yes | Omaha |
| Humboldt | Stream overflow, local drainage | L b |  | No | Yes | Rock Island |
| Independence | Stream overflow | L b<br>CSR a<br>L b | CE e 62    2 | No | Yes | Rock Island |
| Indianola | None |  |  | No | No | Rock Island |
| Iowa City | Stream overflow | CSR ab<br>S ab | CE a       2<br>CE e 66 | No | Yes | Rock Island |
| Iowa Falls | Stream overflow | CSR b |           2 | No | No | Rock Island |
| Jefferson | None | L b<br>L b |  | No | No | Rock Island |

260

| | | | | | |
|---|---|---|---|---|---|
| Keokuk | Stream overflow | CSR ab<br>L b | CE e 62 ** 2 | Yes | Rock Island |
| Knoxville | None | L b | | Yes | Rock Island |
| Le Mars | Stream overflow, local drainage | CSR ab<br>L b | CE e 56  2 | No | Omaha |
| Manchester | Stream overflow | CSR c<br>L b | 2 | No | Rock Island |
| Maquoketa | Stream overflow | CSR c<br>L b<br>S a | L a  2 | No | Rock Island |
| Marion | Stream overflow | | 2 | No | Rock Island |
| Marshalltown | Stream overflow | CSR a<br>L b | CE b<br>L a  2 | Yes | Rock Island |
| Mason City | Stream overflow, local drainage | CSR ab<br>L b | CE e 62<br>L a  2 | Yes | Rock Island |
| Missouri Valley | Stream overflow, local drainage | CSR ab | CE e 43  2 | No | Omaha |
| Monticello | Stream overflow | | 2 | No | Rock Island |
| Mount Pleasant | None | L b | | No | Rock Island |
| Mount Vernon | None | | | Yes | Rock Island |
| Muscatine | Stream overflow | CSR ab<br>L b | CE a  2 | Yes | Rock Island |
| Nevada | None | | | No | Rock Island |
| New Hampton | Stream overflow | L b<br>L b | 2 | No | Rock Island |
| Newton | None | | | Yes | Rock Island |
| Oelwein | Stream overflow | L b | 2 | No | Rock Island |
| Onawa | Stream overflow | | 2 | No | Omaha |
| Orange City | None | L b<br>L b | | Yes | Omaha |
| Osage | None | | | No | Rock Island |
| Osceola | None | L b | | No | Rock Island |
| Oskaloosa | None | L b | | No | Rock Island |
| Ottumwa | Stream overflow, local drainage | CSR ab<br>L b | CE b<br>L a  2 | Yes | Rock Island |

|  | | Status of Flood | | | | |
|---|---|---|---|---|---|---|
| Place | Type of Flood Problem | Studies | Protection | | Maps | Flood Warning | Information Sources |
| Pella | None | | | | No | No | Rock Island |
| Perry | Stream overflow | CSR a | CE a | 2 | No | No | Rock Island |
| Red Oak | Stream overflow, local drainage | SCS a / L b | SCS e 62 | 2 | No | Yes | Omaha |
| Rock Rapids | Stream overflow | L b | | 2 | No | Yes | Omaha |
| Sac City | Stream overflow | CSR b | | 2 | No | No | Rock Island |
| Sheldon | Stream overflow | L b / CSR b / L b | CE e 55 | 2 | Yes | Yes | Omaha |
| Shenandoah | Stream overflow, local drainage | CSR ab / L b | CE e 43 | 2 | No | Yes | Omaha |
| Sibley | Stream overflow, local drainage | CSR ab | | 2 | No | No | Omaha |
| Sioux City | Stream overflow, local drainage | SCS c | CE a | 2 | Yes | Yes | Omaha |
| Spencer | Stream overflow, local drainage | CSR b | | 2 | Yes | Yes | Omaha |
| Spirit Lake | Local drainage | | | 2 | No | Yes | Omaha |
| Storm Lake | None | L b | | | Yes | Yes | Rock Island |
| Tama | Stream overflow | CSR ab / L b | CE e 62 | 2 | No | No | Rock Island |
| Tipton | None | | | | Yes | No | Rock Island |
| Toledo | None | L b | | | No | No | Rock Island |
| Urbandale | Stream overflow | L b | | 2 | Yes | No | Rock Island |
| Vinton | Stream overflow | CSR b | | 2 | No | No | Rock Island |
| Washington | None | | | | No | No | Rock Island |
| Waterloo | Stream overflow | CSR b / L b | CE c / L a | 2 | Yes | Yes | Rock Island |
| Waukon | None | | | | No | No | St. Paul |
| Waverly | Stream overflow | CSR b | | 2 | Yes | Yes | Rock Island |

| | | | | | |
|---|---|---|---|---|---|
| Webster City | Stream overflow | CSR b<br>L b | 1 | No | Yes | Rock Island |
| West Burlington | None | L b | | Yes | No | Rock Island |
| West Des Moines | Stream overflow, local drainage | CSR b<br>S b<br>L b | 2 | Yes | No | Rock Island |
| West Union | None | | | No | No | Rock Island |
| Windsor Heights | Stream overflow | CSR b | 2 | Yes | Yes | Rock Island |
| Winterset | None | | | No | No | Rock Island |

# INDEX

Abstracts, property, 143
Acquisition of flooded areas, 66–67, 105, 116, 138–39, 242
Adams County, 81
Agricultural land use, 4, 14, 87, 90, 152
American Institute of Planners, 18
American Red Cross, 64
American Society of Civil Engineers, 18, 85
Arno River, Italy, 53, 54
Attitudes, 57–58

Backwater effects, 46–47, 224, 228
Beneficial uses of water, 97–98
Benefits, 15, 90, 133, 134
Bi-State Metropolitan Planning Commission, 93, 99–101
Black Hawk Creek, Davenport, 105
Black Hawk Creek, Waterloo, 61, 126
Black Hawk Metropolitan Planning Commission, 93
Blair, D. J., 167
Bucket surveys, 38
Building codes, 67, 82, 191
Building permits, 107
Bypasses, 68, 71

Capital improvement programs, 82, 96, 99, 115
Catfish Creek, 32
Cedar River, 61, 140
  Cedar Rapids program, 137–46, 213
  Waterloo program, 123–35
Certificate of approval, 107
Channel improvements. See Flood protection works
Citizens' advisory committee, 101

City council, 79, 81, 185
Commercial uses, 58–59
Common-law rule, 168
Compensation for taking land, 169, 250–51
Comprehensive planning, 82, 85, 97–99, 151, 160, 162, 176, 189, 225, 238, 241
Comprehensive river basin planning, 12, 16, 47, 54, 67, 89–99, 162, 244, 248
Condemnation, 170, 202
Confiscatory regulations, 5, 169
Cooper, J. F., 219
Cooperative efforts, 6, 17, 21, 68, 96, 129, 142–43, 184, 185, 197, 208, 250
Coralville Reservoir, 5, 6, 7, 71, 216, 219
Council of State Governments, 18, 217
Councils of government, 96
County board of supervisors, 79, 81, 101, 185
County conservation board, 144
Crest stage gages, 202

Dalrymple method, 44
Damage, flood
  annual losses, 3, 13, 68, 86, 157–58, 167, 184, 198
  in Iowa, 31, 34, 39–40, 65–66, 124
  potential, 4, 133
  prevention programs, 16, 215
  reduction of, 71, 103, 215
  residential, 48
  surveys, 38
Datum, river, 63
Davids Creek, 40, 41
Delaware River Basin Commission, 18
Demonstration Cities and Metropolitan Development Act of 1965, 95

**265**

Design flood discharge, 42, 45, 164, 226
Des Moines River, 33, 40, 47, 61, 63, 67, 201
Development of flood plains, 4, 13–14, 48, 58–59, 88, 90, 108–9, 119–20, 129, 139–42, 152–53, 158, 178, 185, 221, 241, 251
Development rights, 178, 192
Devil Creek, 30
Disaster aid law, 66
Disaster area, 65–66
Disease, 171
Districts
  flood channel, 109
  special purpose, 95, 160
  zoning, 175, 187–88
Dougal, M. D., 7, 53
Drainage areas, 34, 41, 204
Dry Creek, 143, 213, 225, 227
Dry Run Creek, 71
Duck Creek, 45, 104, 213
Due process, 169

Easements, 192, 242
East Fork 102 River, 213
East Nishnabotna River, 61
Economics of flood plain use, 4, 13, 34, 41, 44, 56, 67–68, 81, 89, 110, 134, 138–39, 171–72, 185, 204
Elevation controls, 7, 8, 106, 108, 187–88, 222–23, 227
Ellis, D. W., 197
Emergency measures. See Flood fighting
Enabling legislation, 169–77
  need for, 172, 189–93
Encroachment limits, 7, 46, 109, 164, 177, 219, 221, 223, 227
Engineering works. See Flood protection works
Envelope curves, 32, 45, 46
Equal protection concept, 168
Erosion, 39, 71
Executive order, 18, 113, 208
Extraterritorial zoning control, 190

Farmers Home Administration, 95
Federal Aid Highway Act of 1962, 95
Federal government, 16, 18–20
Federal Housing Act of 1954, 152
Flash flood warning network, 61–64
Flood. See also Damage, flood
  annual peak discharge, 28
  characteristics, 33, 86, 164
  defined, 85
  design, 42–43, 108, 164, 188, 221, 226
  envelope curves, 32–33, 45–46
  hazard, 57–59, 113–14, 157
  histogram, 200
  intermediate regional, 43
  magnitude and frequency, 28, 43, 54, 106–7, 199, 224
  mean annual, 28, 34, 43–44
  peak discharge, 28, 39, 45
  potential, 7, 32–34, 37, 41, 54, 106, 188, 229
  probability, 28, 44, 54, 200, 221, 225
  probable maximum, 42, 221
  stage, 63, 227
  standard project, 42, 221, 225
Flood control. See also Flood protection works
  local committees, 126
  methods for, 55, 68
  relation to management, 12–13
  state statutes, 162–63
Flood Control Act of 1960, 208
Flood fighting
  and emergency measures, 64–66, 71, 142
  operating schedule, 65
  responsibility, 64
Flood forecasting and warning, 59–64
  flash flood warning, 61–63
  national program, 59
  river forecast centers, 59, 60
  sequential steps, 60, 61
Flood insurance. See Insurance
Flood plain(s)
  construction control, 163–64
  defined, 67, 85, 163, 177
  delineation, 227
  development, 13, 47
  encroachment, 14, 46, 54–55, 68, 105, 140, 159, 164, 170, 185, 219
  historical role, 4, 13–14
  legislation, 4, 159
  occupancy, 3, 13–14, 48, 66–67, 86, 90–91, 103, 137, 187, 198
  occupancy charges, 15
  urban extent of, 47, 48
  zoning. See Plan and zoning commissions; Zones; Zoning; Zoning control; Zoning ordinances
Flood plain information studies. See Information studies
Flood plain management. See Management, flood plain
Flood plain regulations. See Regulations; Zoning
Flood proofing, 67, 88, 131, 191, 215, 228, 229
Flood protection works, 4, 14, 68–72, 90, 118, 125, 167, 215, 242, 252
  channel improvements, 68–69, 89, 105, 224
  coordinated with urban renewal, 118, 128–29, 242
  dams, 71, 89

# INDEX

diversion channels, 71
engineering studies, 68, 221, 252
expenditures, 3, 4, 13, 68, 89, 118–19, 157, 167, 184
floodwalls, 68–70, 88, 126
levees, 46, 64, 68–71, 89, 126
reservoirs, 68, 71, 89–90
state approval, 163
watershed management, 68, 71, 244–45
Flood routing, 47
Floods
catastrophic, 5, 42, 54, 89, 238
experienced, 3, 28, 37–38, 40, 43, 54, 86, 106, 124, 126, 159, 199, 200, 238
flash, 4, 5, 38, 40
regional, 7, 34
rescue operations, 4, 64
Floodway, 108, 164, 219
defined, 67, 86, 163, 177
easements, 192
fringe, 86, 177
Floyd River, 5, 29, 30
Four-Mile Creek, 61, 201
Functional economic area, 81
Funk, J. W., 37
Future flood problems, 37

Gardner, D. K., 137
General Assembly, 66, 156, 159
General safety and welfare, 171, 173, 174, 176, 186
Goddard, J. E., 11
Greenbelt, 139, 145. *See also* Open-space uses
Gully control, 71, 245
Gumbel method, 44

Hazen method, 44
Health
hazards, 16, 171
regulations, 16, 106, 156, 171–72
Hines, N. W., 167
Housing Act of 1949, 126
Housing codes, 191
Howe, J. W., 3
Hydraulic characteristics, 87
Hydrologic atlases, 18, 199–206. *See also* U.S. Geological Survey
Hydrologic regions, 34, 38, 45
Hydrology, 33, 38, 54, 60, 199

Indemnification of losses, 15, 17
Indian Creek, Cedar Rapids, 62, 140, 213, 225, 227
Indian Creek, Council Bluffs, 5
Industry, 16, 58, 81, 124, 129, 134

Information studies, 20, 57–59, 115, 198, 208, 221, 240, 255
Cedar Rapids, 143
Davenport, 107, 108
Iowa City, 7, 219
Iowa Natural Resources Council, 221
U.S. Corps of Engineers, 207–18
U.S. Geological Survey, 197–206, 240
Insurance
flood, 17, 19, 20, 67
mortgage, 115, 119
Interdisciplinary concern, 11, 12, 16, 19, 20–21, 165, 249
Interior drainage, 68, 71
Inundation limits, 7, 219
Iowa Adjutant General, 64
Iowa City Planning and Zoning Commission, 6, 7
Iowa Code, 80, 160, 162, 173–77, 186–92
Iowa Development Commission
coordination with HUD, 115, 152–53
coordination with INRC, 152–53, 229
state program, 151–55
Iowa Geological Survey, 64
Iowa Highway Patrol, 64
Iowa Highway Research Board, 28
Iowa Natural Resources Council
coordination role
at Cedar Rapids, 143–44, 225, 227
at Davenport, 103, 107–11
with federal FPMS program, 212, 221, 229
at Iowa City, 6–7
with Iowa Development Commission, 152–53, 229
in planning, 107–8, 164–65, 189
with USGS flood mapping program, 221, 229
design flood, 45–46, 226
establishment, 159
flash flood warning system, 61
flood control, 72
flood plain management program, 165, 219–32
regulation role, 46–47, 83, 106–7, 152, 159–65
specific regulatory powers, 83, 152–53, 163, 177
supplementary rainfall data, 38
zoning coordination, 106–7, 177, 189, 229–30
Iowa River, 8, 32, 61, 219
Iowa State Office of State Planning and Programming, 115
Isohyetal map, 39

Johnson, E. O., 103
Johnson, L. B., President, 2, 18, 208, 216

Johnson County Zoning Commission, 7, 8

Kansas River, Kans., 199
Klatt, W. R., 85

Land management, 54, 71
Land treatment, 71
Land-use planning, 15–16, 66, 85, 96
　background, 79–83
　coordinated with flood protection, 68, 88, 129, 134
　coordination problems, 99, 152
　cost sharing, 152, 197
　districts, 108
　elements, 82, 97, 114
　federal programs, 113–22
　in Iowa, 151–56
　mix of uses, 124
　regional, 93, 97–99
　rotating supply, 91
　state statutes, 80
　surveys, 82, 97
　urban planning assistance, 115
Larson, P. L., 113
Legal nuisance, 168
Legal principles, 4, 6, 158, 159
Leisure time, 139, 144
Little Sioux River, 29, 69
Lizard Creek, 61
Loans, 66, 110, 204, 208
Local flood relations program, 249
Love, S. K., 174
Luhman, W. S., 93

Mad Creek, 5
Management, flood plain, 4, 11, 53, 68
　bibliography, 16
　Cedar Rapids program, 143–44, 225, 227
　coordination problems, 103–11, 106–7, 164, 189
　Davenport program, 103–11
　Iowa objectives, 220–21
　legal context, 167–82, 183–93
　local government role, 17, 19, 21, 109, 186, 237, 244
　meaning, 12, 13
　national program, 13, 18, 20, 94, 113–15, 198
　principles, 54, 55, 73, 123
　program elements, 56, 73, 99, 114
　purpose, 13
　state programs, 17, 19, 20, 107–9, 157, 219–32
　state statutes, 157–66
　summary report, 226

Tennessee Valley Authority, 247–53
　Waterloo program, 124–35
Mapped streets act, 99
Mapping, 82, 189, 197, 199, 223
Maynard, J. L., 183
McLaughlin, W. M., 151
Mean annual flood, 28–34
Miami Conservancy District, 89, 90
Milldams, 160
Minimum elevations. *See* Elevation controls
Minimum protection level, 46–47, 158, 219, 221, 227
Mississippi River, 4, 33, 34, 47, 53, 59, 66, 86, 88, 90, 93, 100, 102, 104, 162, 199, 216
Missouri River, 47, 59, 71, 90, 162
Mobile home parks, 192
Mortgage insurance, 115, 119, 208
Multipurpose river development, 68, 71, 89–90, 101, 248
Municipal corporations, 170–71
Murphy, F. C., 67

Navigation, 97
Nebraska Unicameral, 245
Nishnabotna River, 31
Nonconforming uses, 110, 111, 178, 179, 189
Northeast Illinois Metropolitan Area Planning Commission, 18

Ohio River, 88
Open-space uses, 67, 88, 90, 117, 242
　Cedar Rapids program, 137–46
　criteria, 138, 188
　Davenport program, 108–10
Ordinances, 173–77, 183–93. *See also* Zoning ordinances

Parks, B. A., 79
Park uses, 108, 138, 239. *See also* Open-space uses
Passaic River, New Jersey, 199
Pearson method, 44
Permanent evacuation, 67
Person, defined, 163
Peterson, C. E., 7, 157
Physiographic characteristics, 34, 45
Pine Creek, 31
Plan and zoning commissions
　authority, 81, 176
　comprehensive planning, 82, 176
　duties, 79, 81, 176
　enabling acts, 80, 172, 173
　permissive nature, 80, 250
　regional, 97

## INDEX

Planned-unit developments, 120, 189
Planning. *See* Comprehensive planning; Land-use planning
Plumbing code, 192
Police power, 158, 169, 186
 for Iowa municipalities, 170-71
Population growth in urban areas, 93, 138, 151, 152
Prairie Creek, 62, 140, 213
Precipitation, 38
 magnitude and frequency, 41, 42
 probable maximum, 41, 221
Probable maximum flood, 42, 221
Project efficiency, 47
Public health and safety, 156. *See also* Health
Publicity and public relations, 16, 20, 144, 203, 216, 229, 250
Public works planning, 116

Raccoon River, 31, 201
Rainfall, 29, 33, 38. *See also* Precipitation
Rathbun Reservoir, 71
Recognizing flood hazards, 57
Recreation, 91, 98, 117, 124, 128, 131, 133, 137-46
 classifying uses, 139
Red Rock Reservoir, 71
Regional floods, 7, 34
Regional planning, 80, 93, 96, 115, 154
Regulations, flood plain
 concepts, 12, 67
 coordination problems, 103-11, 142-43, 153, 156, 184-85, 229, 251
 enforcement, 109, 164-65, 202
 for one watercourse of several, 179
 hydrologic factors, 34, 38-45
 model ordinance, 167-82
 need for, 4, 15-16, 56, 152, 155
 nonconforming uses, 178-79, 189
 policy, 171-72
 reasonableness, 168-69, 189
 statutory control, 157-66
 utility extensions, 240
Residential uses, 47-48, 59-59, 81, 152
Resolution of necessity, 239
Restrictive covenants, 190
Rio Grande River, Tex., 53, 217
Rock River, Ill., 100
Rule of construction, 174

Safety, public, 156
Salt Creek, Nebr., 238
Salvation Army, 5, 6, 64
Sanitary landfills, 243
Saylorville Reservoir, 71, 216
Schroeder, J. H., 237

Schwob, H. H., 27
Sediment, 105-6, 131, 245
Sheaffer, J. R., 67, 123, 204
Skunk River, 34, 213
Slackwater area, 86
Snowmelt floods, 33, 42
Social costs and benefits, 15, 67-68, 90
Soil conservation practices, 68, 71, 87, 244
Squaw Creek, 61, 213
Standard metropolitan statistical areas, 94, 151
Standard project flood, 42, 221, 225
State statutory control, 106-7, 155
Stephenson, J. N., 207
Storm drainage, 89, 117, 190
 in Iowa, 29-32, 38, 41, 199
Storm transposition, 38
Storms, severe, 5, 38, 199
Stream classification, 61
Structures, flood plain, 88, 109, 140-43, 164, 178
Subdivision regulations, 67, 82, 99, 190-91, 250
Subsidies, 14
Substantive due process, 169

Tarkio River, 32
Task Force on Federal Flood Control Policy, 13, 17
Tax assessments, relief from, 67
Technical advisory committee, 101
Tennessee River, Tenn., 248
Tennessee Valley Authority, 16, 18, 209, 217, 247
Thunderstorms. *See* Storms, severe
Time lag, 61
Topographic mapping. *See* Mapping
Transportation routes, 88, 152
 parkways in flood plains, 105
 planning, 95, 115

Union Park Creek, 32
U.S. Agricultural Stabilization and Conservation Service, 64
U.S. Army Corps of Engineers, 17, 38, 42, 44, 63, 64, 101, 103, 104, 126, 133, 143, 157, 207
 flood control, 14, 72, 126
 flood description, 42
 FPMS program, 18 107-8, 143-44, 207-18, 221-22, 225, 239, 249
U.S. Army Weapons Command, 101
U.S. Bureau of Budget, 13, 17, 208
U.S. Bureau of Census, 151
U.S. Bureau of Outdoor Recreation, 19, 144
U.S. Bureau of Public Roads, 101

U.S. Bureau of Reclamation, 19
U.S. Constitution, 168
U.S. Department of Housing and Urban Development, 17–18, 95, 109, 113–22, 144, 204, 242, 251
 field operations, 121
 programs, 19, 114
 urban planning assistance, 152
 Urban Renewal Administration, 126
U.S. Environmental Science Services Administration, 19, 44, 53, 59, 61–63, 64
U.S. Geological Survey, 44, 63–64, 221
 flood plain mapping, 18, 197–206
 flood studies, 27, 38, 64
 topographic mapping, 197
U.S. Office of Civil and Defense Mobilization, 64
U.S. Office of Emergency Planning, 17
U.S. Red Cross, 64
U.S. Senate Select Committee, 17
U.S. Soil Conservation Service, 19, 38, 44, 46, 64, 72, 157, 158, 244
U.S. Weather Bureau, 19, 44, 53
 flash flood warning system, 61–63
 flood forecasting, 59–64
Upper Iowa River, 71
Urban growth, 47, 82, 85, 94, 103, 105, 123, 138, 151, 185, 202, 238
Urban redevelopment, 23–35
Urban renewal, 67, 82, 118, 128
Utilities, 105, 152
 control, 82, 88, 204, 239–41
 regulation, 67

Valley storage, 46–47
Veterans Administration, 204
Volga River, 216

Walnut Creek, 61, 201
Wapsinonoc Creek, 31, 38, 41
Wapsipinicon River, 100
Warning signs, 67
Water resources
 development, 20, 94, 97, 162, 197, 248
 management, 89, 162, 199
 strategy, 108–9, 132
Water Resources Council, 44, 45
Water Resources Planning Act of 1965, 17
Watershed
 development, 71, 245
 urbanization, 202
Water supply papers, 198–99
Water surface profiles, 47, 54–55, 108, 200, 221, 224, 227
Wayman Creek, 31
Weather surveillance radar, 61
Weathers, J. W., 247
Weaver, Secretary, 114–15
Weldon River, 31
Winnebago River, 61

Zones
 restrictive, 86
 valley channel, 7
 valley plain, 7
Zoning
 board of adjustment, 187
 commissions, 79–80, 175. See also Plan and zoning commissions
Zoning control, 80, 99, 158, 175, 186
 introduction, 151–56
 relation to management, 12–13, 67, 82
Zoning ordinances, flood plain, 67
 administration, 192
 approval by INRC, 160, 165, 177
 boundaries, 108–9, 177–78
 Davenport, 106, 108–10
 districts, 7, 108, 175
 enactment, 81, 175
 Iowa City, 7, 8
 minimum elevations, 8, 106
 relation to planning, 82
 schedule, 186
 state enabling acts, 80, 159–60, 169–77, 186
 structures permitted, 7, 178
 techniques, 187
 text, 186
 zoning map, 186–88